Flat Space Cosmology

Flat Space Cosmology

A New Model of the Universe Incorporating Astronomical Observations of Black Holes, Dark Energy and Dark Matter

Eugene Terry Tatum

and

U.V.S. Seshavatharam

Universal-Publishers

Irvine • Boca Raton

Flat Space Cosmology: A New Model of the Universe Incorporating Astronomical Observations of Black Holes, Dark Energy and Dark Matter

Universal Publishers, Inc.
Irvine • Boca Raton
USA • 2021
www.Universal-Publishers.com

ISBN: 978-1-62734-339-8 (pbk.)
ISBN: 978-1-62734-340-4 (ebk.)

Typeset by Medlar Publishing Solutions Pvt Ltd, India
Cover design by Ivan Popov

Library of Congress Cataloging-in-Publication Data

Names: Tatum, Eugene Terry, 1955- author. | Seshavatharam, U. V. S., author.
Title: Flat space cosmology : a new model of the universe incorporating
astronomical observations of black holes, dark energy
and dark matter/Eugene Terry Tatum and U.V.S. Seshavatharam.
Description: Irvine, California : Universal Publishers, [2021] |
Includes bibliographical references.
Identifiers: LCCN 2021007590 (print) | LCCN 2021007591 (ebook) |
ISBN 9781627343398 (paperback) | ISBN 9781627343404 (ebook)
Subjects: LCSH: Cosmology. | Black holes (Astronomy) | Dark energy
(Astronomy) | Dark matter (Astronomy)
Classification: LCC QB981 .T385 2021 (print) | LCC QB981 (ebook) |
DDC 523.1--dc23
LC record available at https://lccn.loc.gov/2021007590
LC ebook record available at https://lccn.loc.gov/2021007591

Table of Contents

Preface

When one combines the observational results of the past three decades with those pending in the current decade, it appears that there has never been a better time to be an astrophysicist or cosmologist. Most would agree that we are fortunate to be living in an observational and theoretical 'Golden Age' with respect to understanding our place in the universe.

Over the previous decade, both of us, working at opposite ends of the world, have been involved on the theoretical side as cosmologists. Making use of the internet, we joined forces in 2015 after one of us (E.T.T.) published two papers in *Journal of Cosmology* commenting on initial Planck satellite survey findings. We were both simultaneously struck by something not widely recognized or discussed at the time.

The results of the Planck survey which *were* widely discussed crystalized the suspicion that our observable universe is at, or very near, what is called 'critical density.' To put it simply, the observations indicated that the total observable mass of the universe is a tiny fraction less than what would be necessary to decelerate the universal expansion into a 'Big Crunch' in the distant future. To use the vernacular for 'critical density,' the observable universe is spatially flat to an extraordinary degree (perhaps to within one part in 10^{60}!).

The critical density confirmation was indeed impressive, but what captured our attention more than anything else was the observational implications concerning the ratio of the mass of the observable universe to its radius. One can readily check these numbers for oneself by looking up 'Observable Universe' on Wikipedia. The current best estimate of the mass of *ordinary* matter is 1.5×10^{53} kg and the current best estimate of the radius of the observable universe is 4.4×10^{26} m. If we compare these two numbers in the form of the ratio of the mass M to the radius r (M/r ratio), we get approximately 3.4×10^{26} kg/m for the observable universe. *What is particularly interesting about this number is that it is the same order of magnitude as the M/r ratio of a black hole!* A theoretical Schwarzschild black hole, for instance, has a M/r ratio of $c^2/2G$, which is approximately 6.73×10^{26} kg/m.

Moreover, the M/r ratio comparison becomes even more compelling when we consider that the observable universe has at least 5–6 times the mass of ordinary matter in the form of dark matter. Thus, our observations would imply a total matter mass of about 10^{54} kg within the observable radius of 4.4×10^{26} m. Using the Schwarzschild formula, $r_s = 2GM/c^2$, and plugging in 10^{54} kg for M, we can see that the Schwarzschild radius r_s should be about 1.485×10^{27} m for an object with a mass of 10^{54} kg. One can appreciate the significance of this comparison (4.4×10^{26} m versus 1.485×10^{27} m) by realizing that 'any object whose radius is smaller than its Schwarzschild radius is called a black hole' (as noted in the referenced Wikipedia entry for 'Schwarzschild radius').

Lest we be accused of jumping to unwarranted conclusions, we are not asserting that our universe is a black hole *in the usual sense of the term*. After all, it is clearly expanding rather than collapsing. However, our realization that it actually does have the approximate density of a 13.7 billion light-year radius black hole (9.5×10^{-27} kg/m^3) was the primary inspiration and jumping off point for the publications collected in this book. *The question worth asking is whether or not our universe could be something very much like a time-reversed black hole-like object. If one prefers, such an object can be referred to as a 'white hole'* instead. Furthermore, as discussed in the first chapter, the prior theoretical work of Hawking and Penrose, rigorously *proving* the similar, if not identical, physical nature of astrophysical and cosmological 'singularities,' clearly opens the door for exploring this question. Hawking was one of the first to consider such a question, although, to our knowledge, he never pursued it with a formal cosmological model.

This book offers readers the opportunity to quickly familiarize themselves with the unique cosmological model we first published in 2015. For reasons discussed at length in the summarizing first chapter, our model is called 'Flat Space Cosmology' (FSC) because it models a time-reversed (*i.e.,* expanding) black hole-like object at *perpetual* critical density. We don't believe that current observations of critical density are an inexplicable strange coincidence, but rather an important clue as to how our remarkably fractal-like universe actually works. To use computer programmer jargon, this particular cosmological 'coincidence problem' is not a 'bug' but rather a 'feature.'

The summary first chapter contains the FSC model assumptions and many of its derivations correlating closely with cosmological observations. The second chapter explains why FSC appears to be superior to standard inflationary cosmology in at least eleven different categories. All ensuing chapters are excerpts from identically-entitled, peer-reviewed, FSC papers published in the same chronological order from 2018 to the present. They each address individual key features of the FSC model in logical sequence. The only exceptions are timely chapters 12 and 16, which introduce the reader to a plausible dark matter theory. This leads us to believe that intergalactic dark matter may have been significantly underestimated in terms of its percentage of the critical density. The FSC model and the intriguing dark matter theory are brought together in Chapter 14 to suggest that the total matter mass-energy and dark energy percentages may in fact be closer to 50:50, rather than the present estimate of 32:68. If so, the additional cosmological 'coincidence problem' of similar orders of magnitude for matter energy and dark energy might actually be an important clue as to the nature of dark energy with respect to matter energy. Are matter energy and dark energy within a *scaling* cosmic vacuum possibly two sides of the same cosmological coin? Could a continuous link between cosmic vacuum energy density and matter energy density have something to do with quantum nonlocality through *instantaneous* (*i.e.,* faster than light!) conservation of cosmic energy? We suspect this may be the case in our matter-generating quintessence model. If so, it would not be the first time that matter and energy were found to be deeply interconnected. Significantly, as discussed in this book, the FSC model does not have the well-known vacuum energy density 'cosmological constant problem' that deeply troubles proponents of the concordance model (*i.e.,* standard ΛCDM cosmology).

The final two chapters focus entirely on dark matter and dark energy. These are two of the remaining great mysteries of the universe. The FSC model has something to say about both topics. We trust that you will find both concluding chapters to be interesting!

Eugene Terry Tatum (USA)
U.V.S. Seshavatharam (India)
February, 2021

REFERENCES

Aghanim, N., et al. (2018). Planck 2018 Results VI. Cosmological Parameters. http://arXiv:1807.06209v1

Bennett, C.L. (2013). Nine-Year Wilkinson Microwave Anisotropy Probe (WMAP) Observations: Final Maps and Results. arXiv:1212.5225v3 [astro-ph.CO]. doi:10.1088/0067-0049/208/2/20.

de Bernardis, P., et al. (2000). A Flat Universe from High-Resolution Maps of the Cosmic Microwave Background Radiation. arXiv:astro-ph/0004404v1. https://doi.org/10.1038/35010035

Macaulay, E., et al. (2018). First Cosmological Results Using Type Ia Supernovae from the Dark Energy Survey: Measurement of the Hubble Constant. arXiv:1811.02376v1

Perlmutter, S., et al. (1999). The Supernova Cosmology Project, Measurements of Omega and Lambda from 42 High-Redshift Supernovae. Astrophysical Journal, 517: 565–586. [DOI], [astro-ph/9812133].

Riess, A.G., et al. (1998). Observational Evidence from Supernovae for an Accelerating Universe and a Cosmological Constant. Astronomical Journal, 116(3): 1009–38.

Schmidt, B., et al. (1998). The High-Z Supernova Search: Measuring Cosmic Deceleration and Global Curvature of the Universe Using Type Ia Supernovae. Astrophysical Journal, 507: 46–63.

Tatum, E.T. (2015). Could Our Universe Have Features of a Giant Black Hole? Journal of Cosmology, 25: 13061–13080.

Tatum, E.T. (2015). How a Black Hole Universe Theory Might Resolve Some Cosmological Conundrums. Journal of Cosmology, 25: 13081–13111.

Tatum, E.T., Seshavatharam, U.V.S. and Lakshminarayana, S. (2015). Flat Space Cosmology as an Alternative to ΛCDM Cosmology. Frontiers of Astronomy, Astrophysics and Cosmology, 1(2): 98–104. http://pubs.sciepub.com/faac/1/2/3

Tatum, E.T., Seshavatharam, U.V.S. and Lakshminarayana, S. (2015). The Basics of Flat Space Cosmology. International Journal of Astronomy and Astrophysics, 5: 116–124. http://doi.org/10.4236/ijaa.2015.52015

Tatum, E.T., Seshavatharam, U.V.S. and Lakshminarayana, S. (2015). Thermal Radiation Redshift in Flat Space Cosmology. Journal of Applied Physical Science International, 4(1): 18–26.

Wikipedia contributors. (2020, November 29). Observable universe. In *Wikipedia, The Free Encyclopedia*. Retrieved 16:37, November 30, 2020, from https://en.wikipedia.org/w/index.php?title=Observable_universe&oldid=991379341

Wikipedia contributors. (2020, November 28). Schwarzschild radius. In *Wikipedia, The Free Encyclopedia*. Retrieved 16:41, November 30, 2020, from https://en.wikipedia.org/w/index.php?title=Schwarzschild_radius&oldid=991083385

CHAPTER 1

A Heuristic Model of the Evolving Universe Inspired by Hawking and Penrose

Abstract: A heuristic model of universal expansion is presented which uses, as its founding principle, Stephen Hawking's singularity theorem. All assumptions of this model are intrinsically linked to Hawking's theorem and its implications with respect to the time-symmetric properties of general relativity. This is believed to be the first mathematical model constructed in such a way, and it is remarkably accurate with respect to current astrophysical observations. The model's origin, basic assumptions and selected observational correlations are presented in this chapter, including its accurate derivation of the observed Hubble parameter value.*

Keywords: Flat Space Cosmology; Cosmology Theory; Hubble Parameter; Cosmic Flatness; Cosmic Entropy; Black Holes; $R_h = ct$ Model

1. INTRODUCTION AND BACKGROUND

A heuristic mathematical model of the evolving universe, for the purpose of this chapter, is one which tracks its *global* parameters (Hubble parameter, radius, mass, energy, entropy, average temperature, temperature anisotropy, *etc.*) as a function of cosmic time. For it to be useful, such a model should be consistent with everything we currently observe about the universe as a global object, and extend these parameters indefinitely into

*Originally published on June 24, 2019 by IntechOpen (see Appendix refs).

the past and future. In assembling such a model, it is particularly useful to start with a founding principle on which some or, preferably, all of the starting assumptions can be based. For this particular model, the founding principle is based upon the groundbreaking work of Roger Penrose [1] and Stephen Hawking [2][3] concerning the similar theoretical nature of astrophysical and cosmological singularities. This founding principle is Hawking's singularity theorem.

Hawking's singularity theorem implies that our universe, following time-symmetric properties of general relativity, could be treated mathematically as if it were a cosmological black hole-like object moving *backwards in time* (*i.e., expanding from* a singularity state as opposed to *collapsing to* a singularity state). Unfortunately, although Hawking's theorem was rigorously logical, he never actually put together a predictive mathematical cosmological model based upon his theorem. *What is presented in this chapter is believed to be the first such model.*

This author (E.T.T.) was sufficiently intrigued by the initial Planck satellite survey results and the potential implications of Hawking's singularity theorem that he teamed up with two Indian physicists (U.V.S. Seshavatharam and S. Lakshminarayana) in 2015 to publish the seminal papers [4][5][6] on this model. For reasons to be discussed below, this model is called 'Flat Space Cosmology' (FSC). The current five basic assumptions of FSC are presented below.

2. FIVE BASIC ASSUMPTIONS OF FLAT SPACE COSMOLOGY

1. The cosmic model is an ever-expanding sphere such that the cosmic horizon always translates at speed of light c with respect to its geometric center at all times t. The observer is operationally-defined to be at this geometric center at all times t.
2. The cosmic radius R_t and total mass M_t follow the Schwarzschild formula $R_t \cong 2GM_t/c^2$ at all times t.
3. The cosmic Hubble parameter is defined by $H_t \cong c/R_t$ at all times t.
4. Incorporating our cosmological scaling adaptation of Hawking's black hole temperature formula, at any radius R_t, cosmic temperature T_t is inversely proportional to the geometric mean of cosmic total mass

M_t and the Planck mass M_{pl}. R_{pl} is defined as twice the Planck length (*i.e.,* as the Schwarzschild radius of the Planck mass black hole). With subscript t for any time stage of cosmic evolution and subscript pl for the Planck scale epoch, and, incorporating the Schwarzschild relationship between M_t and R_t,

$$\left.\begin{array}{l} k_B T_t \cong \dfrac{\hbar c^3}{8\pi G\sqrt{M_t M_{pl}}} \cong \dfrac{\hbar c}{4\pi\sqrt{R_t R_{pl}}} \\ \left\{\begin{array}{ll} M_t \cong \left(\dfrac{\hbar c^3}{8\pi G k_B T_t}\right)^2 \dfrac{1}{M_{pl}} & (A) \\ R_t \cong \dfrac{1}{R_{pl}}\left(\dfrac{\hbar c}{4\pi k_B}\right)^2 \left(\dfrac{1}{T_t}\right)^2 & (B) \\ R_t T_t^2 \cong \dfrac{1}{R_{pl}}\left(\dfrac{\hbar c}{4\pi k_B}\right)^2 & (C) \\ t \cong \dfrac{R_t}{c} & (D) \end{array}\right. \end{array}\right\} \quad (1)$$

5. Total cosmic entropy follows the Bekenstein-Hawking black hole entropy formula [7][8]:

$$S_t \cong \frac{\pi R_t^2}{L_p^2} \quad (2)$$

The rationale for these basic assumptions is closely tied to Hawking's singularity theorem as it might pertain to a time-reversed Schwarzschild cosmological black hole-like object. From the centrally-located observer's point of view, outwardly-moving photons traveling along geodesics at the cosmic boundary (*i.e.,* the fastest-moving 'particles' of the expansion) are infinitely redshifted and thus define the observational event horizon. Therefore, as given in assumption 3, the truly *global* Hubble parameter value can always be defined as speed of light c divided by the ever-increasing Schwarzschild radius R_t. While the first equation of assumption 4 closely resembles Hawking's black hole temperature formula, it is modified so that cosmological mass scales in

Planck mass units. This is thought to be more appropriate for a scaling cosmological model, as opposed to the relatively static thermodynamics of an astrophysical (*i.e.*, stellar) black hole.

As described in some detail in the seminal FSC papers, the first three assumptions allow for perpetual Friedmann's critical density (*i.e.*, perpetual global spatial flatness) of the expanding FSC cosmological model from its inception. It should be emphasized that these assumptions were not adopted for this particular purpose. However, this unexpected and fortuitous outcome is perhaps the most important feature of this model. By dividing the Schwarzschild mass (defined in terms of cosmic radius R_0) by the spherical volume, and substituting c^2/R_0^2 with H_0^2, Friedmann's critical mass density $\rho_0 = 3H_0^2/8\pi G$ is achieved for *any* given moment of theoretical observation (note the subscript '0') in cosmic time. So, *perpetual Friedmann's critical density and global spatial flatness from inception is a fundamental feature of the FSC model. Our model was named for this important feature.*

This perpetual spatial flatness feature, as well as the finite properties of light-speed expansion of the cosmic horizon, obviates the need for an inflationary solution to the cosmological 'flatness problem' and the 'horizon problem.' It also avoids the disturbing and incredible 'infinite multiverse' implications inherent within inflationary cosmic models. The problems of the required new physics of the 'inflaton' field, and of the 'past-incomplete' nature [9] of inflationary models, are also avoided in the FSC model. Many of these differentiating features of FSC with respect to standard inflationary models were discussed at length in a recent FSC summary paper [10] (see Chapter two).

Based upon the relations proceeding from the top equation of assumption 4, and the model Hubble parameter definition of assumption 3, the following FSC log graph can be presented in **Figure 1**.

An overlay of cosmic epochs evolving from the Planck scale epoch, as believed to be the case from particle physics experiments and quantum field theory, is presented in **Figure 2**.

In both figures, there is a tight correlation between cosmic temperature and time elapsed since the Planck scale epoch (not shown) at approximately the 10^{-43} second mark of cosmic expansion.

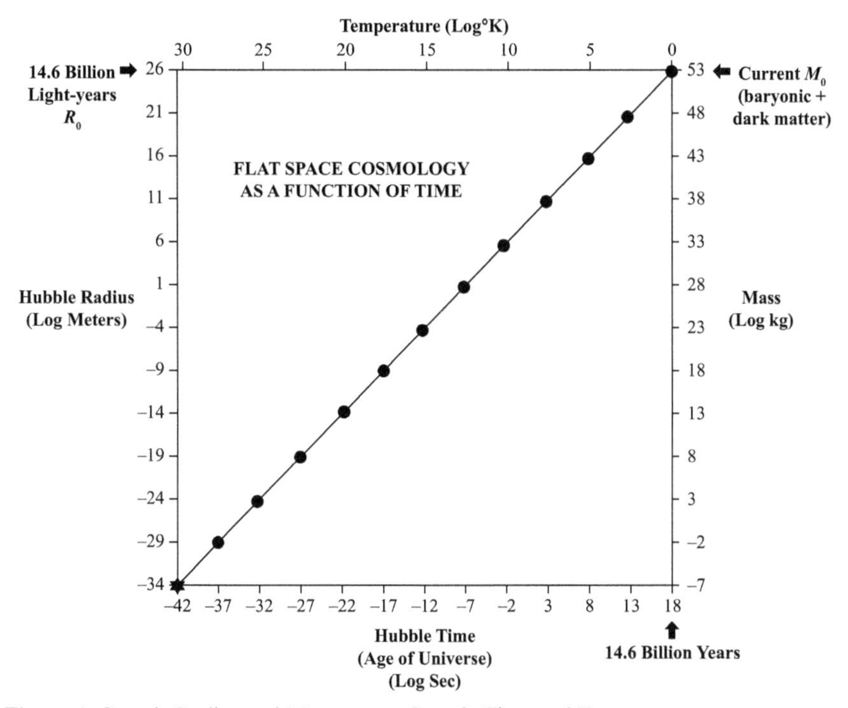

Figure 1 Cosmic Radius and Mass versus Cosmic Time and Temperature.

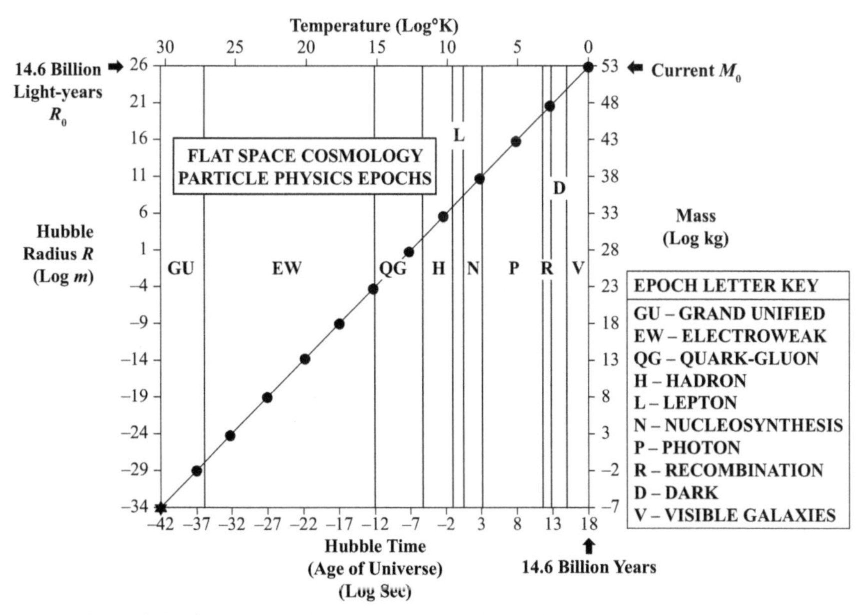

Figure 2 Particle Physics Epochs as a Function of Cosmic Temperature.

3. FSC CORRELATIONS TO ASTRONOMICAL OBSERVATIONS

The following temperature-dependent cosmological parameters can be easily calculated in the FSC model. The only free parameter in any of these equations is the cosmic temperature. Furthermore, by incorporating the values of T_0, $\hbar$, c, G, k_B, and π to as many decimal places as known, any of these FSC parameters can be shown to closely match astronomical observations.

$$R \cong \frac{\hbar^{3/2} c^{7/2}}{32\pi^2 k_B^2 T^2 G^{1/2}} \qquad R_0 \cong \frac{\hbar^{3/2} c^{7/2}}{32\pi^2 k_B^2 T_0^2 G^{1/2}} \tag{3}$$

$$H \cong \frac{32\pi^2 k_B^2 T^2 G^{1/2}}{\hbar^{3/2} c^{5/2}} \qquad H_0 \cong \frac{32\pi^2 k_B^2 T_0^2 G^{1/2}}{\hbar^{3/2} c^{5/2}} \tag{4}$$

$$t \cong \frac{\hbar^{3/2} c^{5/2}}{32\pi^2 k_B^2 T^2 G^{1/2}} \qquad t_0 \cong \frac{\hbar^{3/2} c^{5/2}}{32\pi^2 k_B^2 T_0^2 G^{1/2}} \tag{5}$$

$$M \cong \frac{\hbar^{3/2} c^{11/2}}{64\pi^2 k_B^2 T^2 G^{3/2}} \qquad M_0 \cong \frac{\hbar^{3/2} c^{11/2}}{64\pi^2 k_B^2 T_0^2 G^{3/2}} \tag{6}$$

$$Mc^2 \cong \frac{\hbar^{3/2} c^{15/2}}{64\pi^2 k_B^2 T^2 G^{3/2}} \qquad M_0 c^2 \cong \frac{\hbar^{3/2} c^{15/2}}{64\pi^2 k_B^2 T_0^2 G^{3/2}} \tag{7}$$

Current parameters are calculated in the right-hand column, which incorporates the currently-observed T_0 value of 2.72548 K. Accordingly, the theoretical current FSC Hubble parameter value at this temperature is:

$$H_0 = 2.167862848658891 \times 10^{-18}\ \text{s}^{-1}\ (66.89325791854758\ \text{km.s}^{-1}.\text{Mpc}^{-1})$$

This derived theoretical global H_0 value fits the 2018 Planck Collaboration observational global H_0 value of 67.36 +/– 0.54 km.s^{-1}.Mpc^{-1} (68% confidence interval for TT, TE, EE + lowE + lensing) [11] and the DES 2018 H_0 value of 67.77 +/– 1.30 km.s^{-1}.Mpc^{-1} (SN + BAO) [12]. Since the Planck observational value was obtained with the aid of extraordinarily precise observations of the CMB black body radiation spectrum, this may be as close as we can come in the foreseeable future to a truly *global* Hubble parameter measurement. And yet, the above theoretical H_0 calculation is based *solely* upon this one carefully-measured parameter: $T_0 = 2.72548$ K. This is a remarkable result!

Therefore, one should have great confidence that the following cosmological parameters incorporating the FSC-derived H_0 value are also highly accurate.

$$t_0 \cong \frac{1}{H_0} = 4.61283794 \times 10^{17}\,\text{s}\ (14.61694684 \times 10^9\ \text{sidereal years})$$

(multiplying by 1 sidereal year per 3.155814954×10^7 s)

This value is simply the reciprocal of the above-derived Hubble parameter value, as one would expect for the perpetually spatially-flat FSC cosmic model in comparison to the standard inflationary model. For reasons not elaborated here, *any* inflationary model would be *expected* to calculate a slightly younger cosmic age. 13.8 billion years is now consensus for the standard inflationary model.

$$R_0 \cong \frac{c}{H_0} = 1.38289402 \times 10^{26}\,\text{m}\ (14.617201 \times 10^9\ \text{light-years})$$

(multiplying by 1 Julian light-year per $9.4607304725808 \times 10^{15}$ m)

This current cosmic radius value correlates with current cosmic time by $R_0 = ct_0$. Therefore, FSC is a $R_h = ct$ cosmological model. Later discussion in this chapter will focus on the extremely good statistical fit between $R_h = ct$ models and the accumulated Type Ia supernovae light curve data purported to 'prove' the existence of cosmic acceleration.

$$Vol_0 = \frac{4\pi}{3}\left(\frac{c}{H_0}\right)^3 = 1.10778456 \times 10^{79}\ \text{m}^3$$

$$M_0 = \frac{c^3}{2GH_0} = 9.31126529 \times 10^{52}\ \text{kg}$$

This total mass number can be compared very favorably to a rough estimate made from astronomical observations. The visible matter consists of roughly 100 billion galaxies averaging roughly 100 billion stars each, of average star mass equal to roughly 1.4×10^{30} kg (70 percent of solar mass), totaling to roughly 1.4×10^{52} kg. The 2015 Planck Collaboration report indicates a universal matter ratio of approximately 5.5 parts dark matter to one part visible (baryonic) matter. This brings the total estimated matter in the observable universe to approximately 9.1×10^{52} kg. A recent study [13] of average mass density of intergalactic dust gives a value of approximately 10^{-30} kg.m^{-3}. Since this is approximately one part intergalactic dust to 10^4 parts galactic and peri-galactic matter, intergalactic dust does not appreciably modify the estimated total observational mass of matter given above. Accordingly, this observational estimate is remarkably close to the above FSC theoretical calculation of total cosmic mass attributed to positive energy (*i.e.,* gravitationally attractive) matter.

According to the FSC Friedmann equations (referenced below), the positive matter mass-energy is equal in absolute magnitude, and opposite in sign, to the negative (dark) energy at all times. This is a 50/50 percentage ratio as opposed to the approximately 30/70 ratio implied by yet unproven, and supposedly dark energy-dominating, cosmic *acceleration*. However, without definitively proving cosmic acceleration, standard inflationary cosmology cannot claim this 30/70 ratio! (Please see the discussion and relevant references in the last two paragraphs of this section).

$$M_0 c^2 = \frac{c^5}{2GH_0} = 8.3685479 \times 10^{69} \text{ J}$$

$$\rho_0 = \frac{3H_0^2}{8\pi G} = 8.40530333 \times 10^{-27} \text{ kg.m}^{-3} \text{ (critical mass density)}$$

This closely approximates the observational cosmic mass density calculation of critical density.

$$\rho_0 c^2 = \frac{3H_0^2 c^2}{8\pi G} = 7.554309896 \times 10^{-10} \text{ J.m}^{-3} \text{ (critical mass-energy density)}$$

This closely approximates the observational cosmic mass-energy density and the observational vacuum energy density. They are equal in absolute magnitude, and opposite in sign, in FSC.

A recent paper [14] has integrated the FSC model into the Friedmann equations containing a Lambda Λ cosmological term. Thus, FSC has been shown to be a scalar dynamic Λ dark energy model of the *w*CDM type, wherein equation of state term *w* is always equal to –1.0 (See Chapter 11). Furthermore, it is well-known that a sufficiently realistic $R_h = ct$ model, such as FSC, can fit within the tightest constraints of the Supernova Cosmology Project (SCP) data. Figure 5 of the following link from the SCP is offered as proof: https://dx.doi.org/10.1088/0004-637x/746/1/85 [15]. One can readily see (by the "Flat" line intersection) that a realistic spatially-flat universe model such as FSC is an excellent fit with all such SCP observations to date.

Currently, there is no certainty about the percentage of the critical density which is attributable to dark matter. Those with knowledge of the observational studies of the ratio of dark matter to visible matter realize the difficulty of determining a precise *co-moving* value for this ratio at the present time. Galactic and peri-galactic distributions of dark matter can be surprisingly variable, as evidenced by the 29 March 2018 report in *Nature* [16] of a galaxy apparently completely lacking in dark matter! Although the 2015 Planck Collaboration consensus is a large-scale approximate ratio of 5.5 parts dark matter to one part visible matter, this can only be considered as a rough estimate of the actual *co-moving* ratio, particularly if this ratio varies significantly over cosmic time. A 9.2-to-1 actual ratio in approximately co-moving galaxies (*i.e.,* those within about one hundred million light-years of the Milky Way galaxy) remains a possibility, and would change the ratio of total matter mass-energy to dark energy to essentially unity (*i.e.,* 50% matter mass-energy and 50% dark energy). Thus, the intersection zone of tightest constraints shown in the above-mentioned SCP figure should then correlate with 0.5 Ω_m and 0.5 Ω_Λ. This is one of several important testable predictions discriminating the FSC model from the standard inflationary cosmology model. Precise dark matter measurements of approximately co-moving galaxies are in order, for comparison with the CMB observational Planck Collaboration result.

The question of dark energy density dominance over total matter energy density remains in doubt, at the present time, in the scientific literature. Several recent papers [17][18][19][20][21] have clearly shown that cosmic acceleration, as opposed to the cosmic coasting of $R_h = ct$ models, is not yet *proven*. These are not, of course, refutations of the existence of dark energy as may be defined by general relativity. Rather, they are statistical analyses placing some doubt on dark energy *dominance*, and thus cosmic *acceleration*. These papers are well worth reading.

4. SUPERIORITY OF FSC OVER INFLATIONARY COSMOLOGY

As detailed in the recent FSC summary paper [10] (see Chapter two), there are at least eleven categories in which FSC appears to be superior to standard inflationary cosmology. What makes FSC so powerful in this regard is its ability to make very specific predictions for observations which can be used to falsify the theory if FSC is incorrect. To date, FSC as a global parameter observational predictor has not been falsified.

Standard inflationary cosmology, on the other hand, has largely been cobbled together from observations, and would be difficult to falsify because it makes few falsifiable predictions. The reader should remember that the various theories of cosmic inflation contained *ad hoc* adjustments to accommodate observations [22][23], and that the presumed 'inflaton' energy field of inflation was invented before the actual cosmological vacuum energy now called dark energy was discovered approximately two decades later. It is notable that, rather than attempt to apply the newly-discovered dark energy as a scalar quantity also at work in the early universe, standard inflationary cosmologists have generally assumed the dark energy field to be something entirely distinct from their theoretical inflaton energy field. There has also been an assumption that the post-inflationary energy density of the vacuum must have been a *constant* over the great span of cosmological time. And yet, the theoretical/observational discrepancies created by this 'cosmological constant problem' [24][25] are considered by many to be the most embarrassing problem in all of physics.

The reader is referred to Chapter two for a detailed description as to why FSC is superior to standard inflationary cosmology in the following eleven categories: Cosmic Dawn Early Surprises; Predictions Pertaining to Primordial Gravity Waves; Predicting the Magnitude of CMB Temperature Anisotropy; Predicting the Hubble Parameter Value; Quantifiable Entropy and the Entropic Arrow of Time; Clues to the Nature of Gravity, Dark Energy and Dark Matter; Cosmological

Constant Problem; Dark Matter and Dark Energy Quantitation; Quantum Cosmology; Predicting the Value of Equation of State Term *w*; Requirements for New Physics.

5. SUMMARY AND CONCLUSIONS

This chapter has introduced the reader to the heuristic FSC cosmology model. Like all useful heuristics, FSC provides a means for accurately calculating a variety of parameters. The founding principle for the construction of this model is Hawking's singularity theorem. Accordingly, all assumptions of this model are intrinsically linked to Hawking's theorem and its implications with respect to widely-accepted time-symmetric properties of general relativity. Black holes and black hole-like objects are now known to exist. Furthermore, we know that such objects range over a remarkably wide, fractal-like, scale. Our universe may simply be the largest of these objects which can be observed, albeit from the inside!

Beginning with the work of Penrose and Hawking, the black hole-like properties of the universe have continued to fascinate and surprise us. Our current 'Golden Age' of astrophysical observations and new theories certainly promises even more surprises ahead.

REFERENCES

[1] Penrose, R. (1965). Gravitational Collapse and Space-time Singularities. Phys. Rev. Lett., 14(3): 57–59.

[2] Hawking, S. (1966). Properties of Expanding Universes. Cambridge University Library, PhD.5437. https://cudl.lib.cam.ac.uk/view/MS-PHD-05437/134

[3] Hawking, S. and Penrose, R. (1970). The Singularities of Gravitational Collapse and Cosmology. Proc. Roy. Soc. Lond. A, 314: 529–548.

[4] Tatum, E.T., Seshavatharam, U.V.S. and Lakshminarayana, S. (2015). The Basics of Flat Space Cosmology. International Journal of Astronomy and Astrophysics, 5: 116–124. http://dx.doi.org/10.4236/ijaa.2015.52015

[5] Tatum, E.T., Seshavatharam, U.V.S. and Lakshminarayana, S. (2015). Thermal Radiation Redshift in Flat Space Cosmology. Journal of Applied Physical Science International, 4(1): 18–26.

[6] Tatum, E.T., Seshavatharam, U.V.S. and Lakshminarayana, S. (2015). Flat Space Cosmology as an Alternative to LCDM Cosmology. Frontiers of Astronomy, Astrophysics and Cosmology, 1(2): 98–104. http://pubs.sciepub.com/faac/1/2/3 doi:10.12691/faac-1-2-3

[7] Bekenstein, J.D. (1974). Generalized Second Law of Thermodynamics in Black Hole Physics. Phys. Rev. D, 9: 3292–3300. doi: 10.1103/PhysRevD.9.3292.

[8] Hawking, S. (1976). Black Holes and Thermodynamics. Physical Review D, 13(2): 191–197.

[9] Borde, A., et al. (2003). Inflationary Spacetimes Are Not Past-Complete. Phys. Rev. Lett., 90 (15): 151301. arXiv:gr-qc/0110012.

[10] Tatum, E.T. (2018). Why Flat Space Cosmology Is Superior to Standard Inflationary Cosmology, Journal of Modern Physics, 9: 1867–1882. https://doi.org/10.4236/jmp.2018.910118

[11] Aghanim, N., et al. (2018). Planck 2018 Results VI. Cosmological Parameters. http://arXiv:1807.06209v1

[12] Macaulay, E., et al. (2018). First Cosmological Results Using Type Ia Supernovae from the Dark Energy Survey: Measurement of the Hubble Constant. arXiv:1811.02376v1

[13] Inoue, A.K. (2004). Amount of Intergalactic dust: Constraints from Distant Supernovae and the Thermal History of the Intergalactic Medium. Mon. Not. Roy. Astron. Soc., 350(2): 729–744. https://doi.org/10.1111/j.1365-2966.2004.07686.x

[14] Tatum, E.T. and Seshavatharam, U.V.S. (2018). Flat Space Cosmology as a Model of Light Speed Cosmic Expansion—Implications for the Vacuum Energy Density. Journal of Modern Physics, 9: 2008–2020. https://doi.org/10.4236/jmp.2018.910126

[15] Suzuki, N., et al. (2011). The Hubble Space Telescope Cluster Supernovae Survey: V. Improving the Dark Energy Constraints Above Z>1 and Building an Early-Type-Hosted Supernova Sample. arXiv.org/abs/1105.3470.

[16] van Dokkum, P., et al. (2018). A Galaxy Lacking Dark Matter. Nature, 555: 29–632. doi: 10.1038/nature25767.

[17] Melia, F. (2012). Fitting the Union 2.1 SN Sample with the Rh=ct Universe. Astronomical Journal, 144: arXiv:1206.6289 [astro-ph.CO]

[18] Nielsen, J.T., et al. (2015). Marginal Evidence for Cosmic Acceleration from Type Ia Supernovae. Scientific Reports, 6: Article number 35596. doi: 10.1038/srep35596. arXiv:1506.01354v1.

[19] Jun-Jie Wei, et al. (2015). A Comparative Analysis of the Supernova Legacy Survey Sample with ΛCDM and the R_h=ct Universe. Astronomical Journal, 149: 102.

[20] Tutusaus, I., et al. (2017). Is Cosmic Acceleration Proven by Local Cosmological Probes? Astronomy & Astrophysics, 602_A73. arXiv:1706.05036v1 [astro-ph.CO].

[21] Dam, L.H., et al. (2017). Apparent Cosmic Acceleration from Type Ia Supernovae. Mon. Not. Roy. Astron. Soc. arXiv:1706.07236v2 [astro-ph.CO]

[22] Guth, Alan. *The Inflationary Universe: The Quest for a New Theory of Cosmic Origins*. New York: Basic Books, 1997. (page 238).

[23] Steinhardt, P.J. (2011). The Inflation Debate: Is the Theory at the Heart of Modern Cosmology Deeply Flawed? Scientific American, 304(4): 18–25.

[24] Weinberg, S. (1989). The Cosmological Constant Problem. Rev. Mod. Phys., 61: 1–23.

[25] Carroll, S. (2001). The Cosmological Constant. Living Rev. Relativity, 4: 5–56.

CHAPTER 2

Why Flat Space Cosmology is Superior to Standard Inflationary Cosmology

Abstract: Following recent Cosmic Microwave Background (CMB) observations of global spatial flatness, only two types of viable cosmological models remain: inflationary models which almost instantaneously attain cosmic flatness following the Big Bang; and non-inflationary models which are spatially flat from inception. Flat Space Cosmology (FSC) is the latter type of cosmological model by virtue of assumptions corresponding to the Hawking-Penrose implication that a universe expanding from a singularity could be modeled like a time-reversed black hole. Since current inflationary models have been criticized for their lack of falsifiability, the numerous falsifiable predictions and key features of the FSC model are herein contrasted with standard inflationary cosmology. For the reasons given, the FSC model is shown to be superior to standard cosmology in the following eleven categories: Cosmic Dawn Early Surprises; Predictions Pertaining to Primordial Gravity Waves; Predicting the Magnitude of CMB Temperature Anisotropy; Predicting the Hubble Parameter Value; Quantifiable Entropy and the Entropic Arrow of Time; Clues to the Nature of Gravity, Dark Energy and Dark Matter; Cosmological Constant Problem; Dark Matter and Dark Energy Quantitation; Quantum Cosmology; Predicting the Value of Equation of State Term *w*; Requirements for New Physics.*

---

*Originally published on September 3, 2018 in Journal of Modern Physics (see Appendix refs).

Keywords: Cosmology Theory; Cosmic Inflation; Dark Energy; Cosmic Flatness; CMB Anisotropy; Cosmic Entropy; Emergent Gravity; Black Holes; FSC; Cosmic Dawn; $R_h = ct$ Model

1. INTRODUCTION AND BACKGROUND

The BOOMERanG experiment, Wilkinson Microwave Anisotropy Probe (WMAP) and Planck satellite studies of the Cosmic Microwave Background (CMB) have established beyond reasonable doubt that our universe is spatially flat [1][2][3]. All non-flat cosmologies have thus been consigned to the waste basket of history. Andrew Lange, principle investigator for BOOMERanG, the first study to confirm universal flatness, was decidedly circumspect in his answer to the question on everyone's mind: 'Is this final confirmation that cosmic inflation theory is correct?'

One would do well to remember Lange's admonition that one can never *prove* the correctness of cosmic inflation by confirming spatial flatness in the CMB. He emphasized that there might ultimately be other cosmology theories with a non-inflationary mechanism for achieving CMB flatness. While many scientists at the time considered Lange to be unnecessarily conservative with his answer, others have since reiterated this point of view. Physicist Philip Gibbs, for instance, put it this way: 'The problem... is that no particular model of inflation has been shown to work yet. It is possible that that work has not yet been completed *or that a more recent specific model will be shown to be right*' (our italics) [4].

Given Lange's concerns, physicist Brian Keating and others have set out on a different tack by looking for inflationary B-mode polarization in the CMB. His recent book [Keating (2018)] is an excellent glimpse into this new and exciting phase of experimental cosmology. Unfortunately, no primordial gravity waves of inflation have yet been discovered. In 2014, the BICEP2 team jumped the gun and prematurely announced a confirmatory result, only to be proven wrong by a Planck study overlay of B-mode polarizing interstellar dust.

While more sensitive searches for primordial gravity waves of cosmic inflation are being conducted, other cosmologists have taken up the cause of searching for non-inflationary mechanisms to explain cosmic spatial flatness. Some, including the present authors, have wondered if the universe has always been spatially flat since inception. Certainly, to this point, there is no observational proof otherwise. Even inflationary cosmologists agree that once the universe reaches critical density, it can remain spatially flat for the rest of time. We must remember that inflationary models must achieve spatial flatness to within one part in approximately 10^{60} *before* the first 10^{-32} second of cosmic time for inflation to achieve the degree of flatness now observed in the CMB [5]. This is a crucial realization. The only difference between an inflationary universe and a perpetually flat universe from inception is what happens within that first 10^{-32} second of cosmic time.

Flat Space Cosmology (FSC) is a remarkably accurate cosmological model which was initially developed as a heuristic mathematical model of the Hawking-Penrose implication [6][7] that a universe smoothly expanding from a singularity can be theoretically treated *within the rules of general relativity* as a time-reversed black hole. Thus, it is perhaps not surprising that FSC makes predictions that closely fit with Planck survey observations. For instance, the Planck Collaboration reported an *observed* global Hubble parameter value of 67.8 +/– 0.9 $km.s^{-1}.Mpc^{-1}$ (68% confidence interval) and FSC *predicts* a current global Hubble parameter value of 66.9 $km.s^{-1}.Mpc^{-1}$, which fits the lower end of the Planck survey range. This FSC calculation is based only upon one free parameter measurement, namely Fixsen's fitting of WMAP data for the CMB temperature peak with a black body at 2.72548 K. This CMB temperature number, plugged into FSC equations first published in 2015 [8][9][10][11], allows for a variety of cosmic parameter predictions fitting very tightly with observations. The five assumptions of FSC, closely adhering to the Hawking-Penrose implication, and the FSC observational correlations, were presented in Chapter one.

It is the purpose of this chapter to present the predictions and features of FSC which prove its superiority to standard inflationary cosmology. These predictions and features have been embedded in FSC since its inception in 2015.

2. PREDICTIONS AND SUPERIORITY OF THE FSC MODEL

In terms of observational correlations and/or falsifiability, FSC appears to be superior to standard inflationary cosmology in the following eleven categories:

2.1 COSMIC DAWN AND THE FORMATION OF THE FIRST STARS, QUASARS AND GALAXIES

As noted in several recent papers [12][13][14], standard inflationary cosmology cannot easily explain the surprisingly early formation of the first stars, quasars and galaxies. As detailed in Chapter three [15], temperature curve differences between the two models are such that cosmic dawn, correlating to z redshifts of approximately 15–20, occurred in the FSC model much earlier after the Planck epoch than in standard inflationary cosmology. A comparison of the two temperature curves is shown in **Figure 1**.

The top line is the radiation temperature curve in standard inflationary cosmology and the bottom line is the radiation temperature curve in FSC.

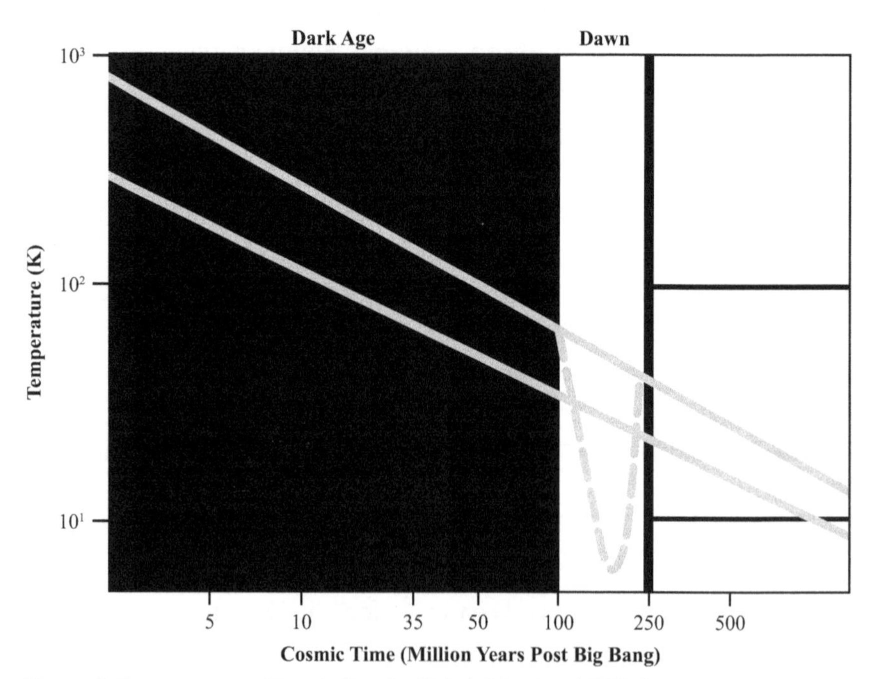

Figure 1 Temperature vs Time in Standard Model (top) and FSC (bottom).

As first presented in Bowman's 2018 publication [16], the dashed curve represents the primordial atomic hydrogen gas temperature.

One should note how cosmic time differs with respect to a given model's radiation temperature. Judging from these temperature curve differences, cosmic dawn in FSC could have been as early as about 35–65 million years after the Planck epoch as opposed to the standard inflationary cosmology cosmic dawn at about 100–250 million years. The nadir of the primordial hydrogen gas temperature, correlating to a *z* value of about 17, would have been at about 50 million years and about 180 million years, respectively. Thus, FSC, by the relative length (14.6 billion years vs 13.8 billion years) and slope of its temperature curve, allows for considerably more time between the formation of the first stars (at cosmic dawn) and the formation of the first quasars and galaxies.

2.2 PREDICTIONS PERTAINING TO PRIMORDIAL GRAVITY WAVES

FSC is a steadily expanding cosmology model, which would not be expected to produce inflationary B-mode primordial gravity waves. There is nothing 'explosive' about the FSC early universe in comparison to the standard inflationary early universe. Thus, FSC predicts that inflationary B-mode primordial gravity waves will never be detected. Such unequivocal detection of inflationary waves would falsify FSC. The continued failure to detect such waves, if the sensitivity of detection methods can be made sufficiently high, would strongly favor FSC over standard inflationary cosmology.

2.3 PREDICTING THE MAGNITUDE OF CMB TEMPERATURE ANISOTROPY

The angular power spectrum of the CMB clearly fits with a spatially flat universe. As noted following the BOOMERanG Collaboration report [17] of CMB anisotropy observations, their results are 'closely fitting the theoretical predictions for a spatially flat cosmological model with an exactly scale invariant primordial power spectrum for the adiabatic growing mode' [18]. Furthermore, the COBE DMR experiment [19] measured a CMB RMS temperature variation of 18 micro-Kelvins. This translates to a dT/T anisotropy value of (0.000018)/2.725 equal to 0.66×10^{-5} (nearly one part in one hundred thousand). This measurement fits within the range of FSC temperature anisotropy predictions for the beginning and ending of the recombination/decoupling epoch [20]. This result clearly favors FSC. See Chapters six and nine for details.

2.4 PREDICTING THE HUBBLE PARAMETER VALUE

In standard inflationary cosmology, the Hubble parameter value can only be determined by observation. That is to say that there is no theoretical ability within standard cosmology to derive a Hubble parameter value. The FSC model, on the other hand, *predicts* the current global H_0 value to be 66.9 kilometers per second per megaparsec (see Chapter one). This fits the 2018 Planck Collaboration [21] and 2018 DES [22] Hubble parameter values. Therefore, this category strongly favors FSC in comparison to standard inflationary cosmology.

2.5 QUANTIFIABLE ENTROPY AND THE ENTROPIC ARROW OF TIME

One of the problems within the standard inflationary model is in quantifying cosmic entropy. Entropy is typically defined in terms of the total number of possible microstates and the probability of a given set of conditions with respect to that number of microstates. These values are impossible to quantify in an infinite-sized inflationary universe or multiverse. FSC, on the other hand, is a *finite* model with a spherical horizon surface area. And, since the Bekenstein-Hawking definition of black hole entropy applies to the FSC model, values for cosmic entropy can be calculated for any time, temperature or radius of the FSC model. Thus, the 'entropic arrow of time' is clearly defined and quantified in the FSC model. The quantifiable entropy in the FSC model allows for model correlations with cosmic entropy theories, such as those of Roger Penrose [23] and Erik Verlinde. Thus, the entropy rules of FSC potentially allow for falsifiability. This feature favors the FSC model, particularly with respect to Verlinde's 'emergent gravity' theory (see below).

2.6 CLUES TO THE FUNDAMENTAL NATURE OF GRAVITY, DARK ENERGY AND DARK MATTER

The reader is referred to Chapter five with the above title for an in-depth discussion of how cosmic entropy in the FSC model may provide for tantalizing clues with respect to the fundamental nature of gravity [24]. In short, the FSC model appears to be the cosmological model correlate to Verlinde's emergent gravity theory [25][26]. Verlinde's landmark paper from 2011 provides strong theoretical support for gravity being an *emergent property* of cosmic entropy. Chapter five makes a case for the correctness of Verlinde's theory. As discussed, if gravity is an emergent property of cosmic entropy, then one might entertain the possibility that dark energy and dark matter could also be emergent properties of cosmic entropy. For instance, perhaps galactic and peri-galactic features attributed to dark matter (such as plate-like galactic rotation and gravitational lensing) could be an unexpected large-scale effect of the entropy of the known galactic baryonic matter. If this turns out to be the correct interpretation, then gravity, dark energy and dark matter might be as difficult to define at the quantum level as 'quantum consciousness' within two connected neurons.

The recent observations of Brouwer, *et al.* [27] appear to be in support of Verlinde's theory as it pertains to dark matter. The discovery of quantum gravity not at all connected to entropy at the quantum level would falsify Verlinde's theory. At present, standard inflationary cosmology, by virtue of its inability to precisely define cosmic entropy, has no capacity to incorporate Verlinde's theory or any other cosmic entropy theory. This appears to favor FSC, particularly in light of the above-mentioned recent observational findings.

2.7 COSMOLOGICAL CONSTANT PROBLEM

The 'cosmological constant problem' is a longstanding problem in theoretical physics. It underscores standard cosmology's inability to unify general relativity with quantum field theory (QFT). Excellent expositions on this subject have been provided by Weinberg [28] and Carroll [29]. QFT theorists calculate a cosmological constant value which differs from observational measurements of the vacuum energy density by a magnitude of approximately 10^{121}! Suffice it to say, this discrepancy is so large that it is often referred to as the most embarrassing problem in all of theoretical physics.

In standard inflationary cosmology it has been assumed that the post-inflationary energy density of the cosmic vacuum must be constant, rather than scalar, over the remainder of cosmic time. However, general relativity does indeed allow for the vacuum energy density to be a dynamic scalar over time. Cosmological models incorporating scaling vacuum energy density are called 'quintessence' models. FSC is one such model. In FSC, the vacuum energy density scales downward by 121.26 logs of 10 over the cosmic time interval since the Planck mass epoch. Perhaps of even greater interest is that the Bekenstein-Hawking cosmic entropy value scales upward in direct proportion to the expanding surface area of the cosmic horizon. If one were to count the current number of Planck mass radius microstates (tiles with an area of $4L_p^2$) within the area of the FSC horizon, the model indicates this entropy number to be a factor of $10^{121.26}$ greater than the 4π value calculated for the Planck mass epoch. Thus, by its implication of a possible relationship between vacuum energy (*i.e.*, dark energy) and total cosmic entropy (as detailed in Chapter five), FSC offers a possible explanation for the magnitude difference between the Planck epoch vacuum energy density calculated by QFT theorists and today's observed vacuum energy density of approximately 10^{-9} J.m^{-3}. Since the FSC model stipulates these values, and standard inflationary cosmology has no basis for deriving them, the FSC model appears to be superior with respect to potentially resolving the cosmological constant problem.

2.8 DARK MATTER AND DARK ENERGY QUANTITATION

As reported by the Planck Collaboration, the ratio of dark matter to visible (baryonic) matter is observed to be approximately 5.5 parts dark matter to one part visible matter. However, there are already significant differences observed between the dark matter-to-visible matter ratios in the galaxies quite near to us (essentially co-movers) and the above dark matter-to-visible matter ratio determined from Planck CMB observations. *Perhaps this ratio is scalar over the great span of cosmic time. If the co-mover ratio is ultimately found to be approximately 9.2, as predicted by FSC, one can then conclude that total matter energy density at present is equal in absolute magnitude to dark energy density*. This equality of opposite sign energy densities is what one would expect for a spatially-flat universe. Otherwise, if one energy density dominated the other, there should be detectable global spatial curvature corresponding to the dominating energy density. One could, in fact, make a strong case that the spatial flatness of the CMB proves the equality of total matter and dark energy densities at the time of the recombination/decoupling epoch. This should nullify any Planck Collaboration conclusions (such as dark energy dominance) which are obviously contrary to their own observations of spatial flatness.

Despite the fact that FSC and standard inflationary cosmology differ somewhat with respect to the percentages of total matter versus dark energy predicted for the co-moving universe, there is one thing about this energy density partition on which everyone agrees: it is truly remarkable that total matter energy density and dark energy density are *of the same order of magnitude at the present time*. As physicist I. I. Rabi once famously remarked, 'Who ordered that?!' This is often referred to as the 'cosmological coincidence problem.' Standard cosmology simply accepts this coincidence problem with no further explanation or rationale. However, FSC *stipulates* perpetual equality of absolute magnitude of these two energy densities as a requirement for a perpetually spatially-flat universe. One can consider this expectation of energy density equality to be a falsifiable FSC prediction with respect to future measurements of total matter energy density in comparison to dark energy density. *An in-depth statistical analysis of approximate co-movers with the Milky Way should give us a better idea of the dark matter to visible matter ratio in the current epoch.*

With respect to standard cosmology's current belief in cosmic acceleration due to dark energy, the reader is referred to references [30] thru [34]. Cosmic acceleration is clearly not proven at the present time, despite the indisputable presence of dark energy as definable within general relativity. There are relative differences in luminosity distance and angular diameter distance formulae in standard inflationary cosmology and $R_h = ct$ modified Milne-type models (like FSC). Two comparative graphs from Chapter four and FSC reference [35] are shown in **Figure 2** and **Figure 3**.

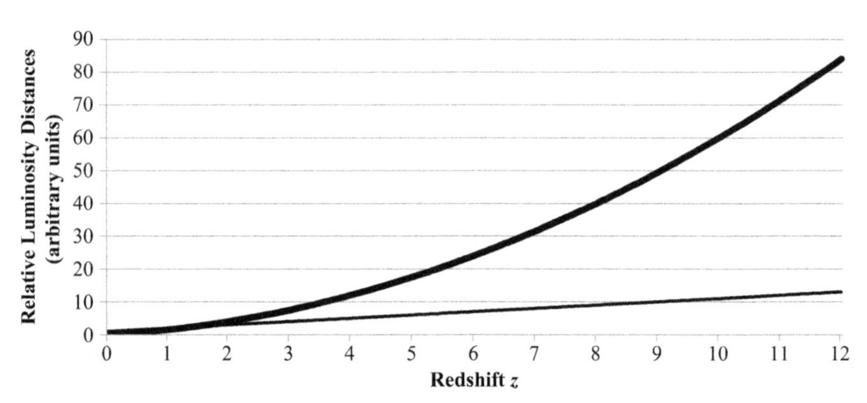

Figure 2 Relative luminosity distances vs. redshift z for standard (thin) and Milne (thick).

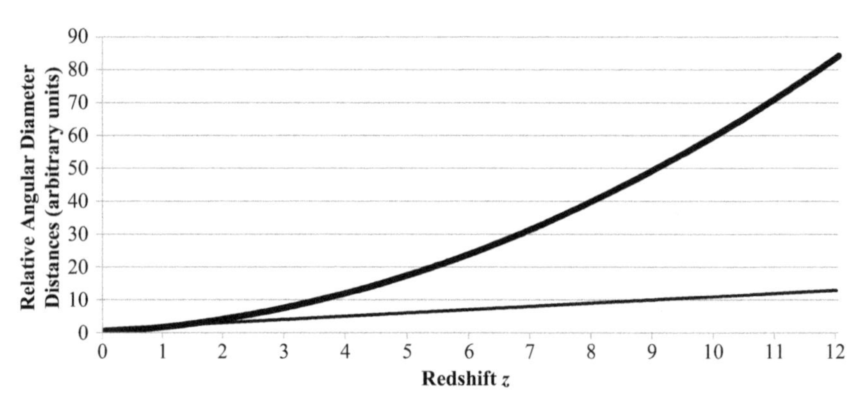

Figure 3 Relative angular diameter distances vs. z for standard (thin) and Milne (thick).

The significance of the relative luminosity distance and relative angular diameter distance comparisons between these two competing models is paramount. An observer of distant Type Ia supernovae expects particular luminosity distances and angular diameter distances to correspond with particular redshifts. If, instead, he or she observes greater-than-expected luminosity distances (*i.e.,* unexpected 'dimming' of the supernovae) or greater-than-expected angular diameter distances, this can easily be misinterpreted by a standard inflationary model proponent as indicative of cosmic acceleration. However, *entirely predictable supernova luminosity distances within a realistic Milne-type universe containing matter, as opposed to a standard model universe, could be one possible explanation for the Type Ia supernovae observations since 1998. Obviously, cosmic acceleration would not then be required to explain these observations.* This possibility, combined with the standard model tension problem presented above (*i.e.*, spatial flatness and dark energy dominance cannot *both* be true at the same time), and the FSC *stipulation* of what standard model proponents refer to as the 'coincidence problem,' strongly favors FSC with respect to its predictions concerning dark matter and dark energy quantitation.

2.9 QUANTUM COSMOLOGY

The FSC model, by virtue of its appropriately scaling cosmic temperature equation (the first equation given in assumption four of Chapter one), can be considered the first successful quantum cosmology model [Tatum, *et al.* (2015)]. By incorporating the values of T_0, $\hbar$, c, G, k_B and π to as many decimal places as known, the FSC parameters can be shown to closely match astronomical observations. For convenience, the most important FSC quantum cosmology equations are repeated here:

$$R \cong \frac{\hbar^{3/2} c^{7/2}}{32\pi^2 k_B^2 T^2 G^{1/2}} \qquad R_0 \cong \frac{\hbar^{3/2} c^{7/2}}{32\pi^2 k_B^2 T_0^2 G^{1/2}}$$

$$H \cong \frac{32\pi^2 k_B^2 T^2 G^{1/2}}{\hbar^{3/2} c^{5/2}} \qquad H_0 \cong \frac{32\pi^2 k_B^2 T_0^2 G^{1/2}}{\hbar^{3/2} c^{5/2}}$$

$$t \cong \frac{\hbar^{3/2} c^{5/2}}{32\pi^2 k_B^2 T^2 G^{1/2}} \qquad t_0 \cong \frac{\hbar^{3/2} c^{5/2}}{32\pi^2 k_B^2 T_0^2 G^{1/2}}$$

$$M \cong \frac{\hbar^{3/2} c^{11/2}}{64\pi^2 k_B^2 T^2 G^{3/2}} \qquad M_0 \cong \frac{\hbar^{3/2} c^{11/2}}{64\pi^2 k_B^2 T_0^2 G^{3/2}}$$

$$Mc^2 \cong \frac{\hbar^{3/2} c^{15/2}}{64\pi^2 k_B^2 T^2 G^{3/2}} \qquad M_0 c^2 \cong \frac{\hbar^{3/2} c^{15/2}}{64\pi^2 k_B^2 T_0^2 G^{3/2}}$$

Current parameters are calculated in the right-hand column. The only free parameter in any of these equations is the cosmic temperature. The right-hand column incorporates the currently-observed Fixsen WMAP T_0 value of 2.72548 K. *It is truly remarkable that, using only current best measurements of the CMB temperature, the derived H_0 value can be calculated and fitted to the lower end of the 2018 Planck Collaboration and DES observational measurements!*

No quantum model exists for standard inflationary cosmology. This obviously favors the FSC model.

2.10 PREDICTING THE VALUE OF EQUATION OF STATE TERM

For reasons given in Chapter 11, the FSC model requires that the equation of state w term have a perpetual value of exactly –1.0. This fits the quantum field theory stipulation that the vacuum pressure p corresponding to the zero-state vacuum energy must always be equal in magnitude to the vacuum energy density ρ (*i.e.*, $p = \rho$). A dark energy dominant universe (as currently believed in standard cosmology) would not meet this quantum field theory stipulation and would have a w value other than exactly –1.0. These are falsifiable predictions for both models. Standard model cosmologists believe our current universe to contain an extremely small *net* negative energy. In other words, they believe in *slight* cosmic acceleration (as opposed to constant velocity light speed expansion), despite current observations of global spatial flatness. However, *if our universe began from a zero-energy state, as is often assumed, and the universe now has a non-zero energy density, however small, this would appear to violate conservation of energy!* Thus, there is tension in the standard cosmology model of dark energy dominance, because the Planck Collaboration has reported $w = -1.006 +/- 0.045$. A final consensus value of exactly –1.0, which seems highly likely at the present time, would be in support of FSC and falsify the current belief in dark energy dominance within standard cosmology.

2.11 REQUIREMENTS FOR NEW PHYSICS

Cosmic inflation theory was invented before the spatial-flattening effects (on positively-curved space-time) of cosmic vacuum energy (dark energy) were discovered in 1998 [36][37][38]. Guth [39] and others [40][41] believed at the time of its invention that a special energy field with inflating features (called by Guth the 'inflaton') was required within the initial 10^{-32} second of universal expansion. It was believed that this energy field was necessary in order to flatten out a presumed highly-curved space-time during and immediately following the inception of expansion. Thus, inflation appeared to be a clever solution to the cosmological 'flatness problem,' as well as the cosmological 'horizon problem.' The latter problem was presumed at the time to exist because most cosmologists believed, without any real evidence, that the universe is infinite and thus otherwise difficult to explain in terms of its remarkable homogeneity in all observational directions. Nevertheless, one of the founders of cosmic inflation, Paul Steinhardt, has pointed out some serious flaws in inflationary theory [42].

For reasons given in Chapter one, FSC solves these cosmological problems without requiring an inflationary epoch. In contrast to inflationary models, in which the total cosmic mass generation is exclusively limited to within a tiny fraction of a second of the Big Bang, *the FSC model is a perpetual mass-generating model* with some similarities to the model presented in the 2019 publication entitled 'A Perpetual Mass-Generating Planckian Universe' by Sapar [43]. This concept of perpetual mass generation has a long tradition going back at least to Hoyle, although Hoyle's particular mass-generating theory was falsified by the discovery of the cosmic microwave background in the 1960s. Here it is important to emphasize that the mystery of mass generation is inherent in *all* cosmology models. FSC simply models perpetual mass generation while inflationary models imply, without any real evidence, that all universal matter was nearly instantaneously created.

It is speculated in Chapters 13 and 14 that negative energy (*i.e.*, gravitationally-repelling energy) within the cosmic vacuum may be continually increasing as an offset to the ongoing production of gravitationally-attracting positive energy in the form of matter. This would be in keeping with the spatial curvature rules of general relativity. One should remember that, according to general relativity, a flat space time is flat precisely because it contains net zero total energy. Furthermore, a globally and perpetually spatially-flat universe which begins from a net zero total energy state (Guth's 'free lunch' idea) would presumably maintain net zero total energy throughout its expansion. Otherwise, a fully self-contained universe, such as a FSC universe, would violate conservation of energy.

Despite the ongoing mystery of mass generation in all cosmology models, for the arguments made above, and for the perpetual mass-generation rationale offered in Dr. Sapar's paper, this category appears to favor FSC in comparison to standard inflationary cosmology.

3. SUMMARY AND CONCLUSIONS

Following recent precise CMB observations of global spatial flatness, only two types of viable cosmological models remain: inflationary models which almost instantaneously attain cosmic flatness following the Big Bang; and non-inflationary models which are spatially flat from inception. FSC is the latter type of cosmological model by virtue of assumptions corresponding to the Hawking-Penrose implication that a universe expanding from a singularity could be modeled like a time-reversed black hole. Since current inflationary models have been criticized for their lack of falsifiability, the numerous falsifiable predictions and key features of the FSC model are herein contrasted with standard inflationary cosmology. For the reasons given, the FSC model is shown in this chapter to be superior to standard cosmology in the following eleven categories: Cosmic Dawn Early Surprises; Predictions Pertaining to Primordial Gravity Waves; Predicting the Magnitude of CMB Temperature Anisotropy; Predicting the Hubble Parameter Value; Quantifiable Entropy and the Entropic Arrow of Time; Clues to the Nature of Gravity, Dark Energy and Dark Matter; Cosmological Constant Problem; Dark Matter and Dark Energy Quantitation; Quantum Cosmology; Predicting the Value of Equation of State Term w; Requirements for New Physics.

REFERENCES

[1] de Bernardis, P., et al. (2000). A Flat Universe from High-Resolution Maps of the Cosmic Microwave Background Radiation. arXiv:astro-ph/0004404v1. https://doi.org/10.1038/35010035

[2] Bennett, C.L. (2013). Nine-Year Wilkinson Microwave Anisotropy Probe (WMAP) Observations: Final Maps and Results. arXiv:1212.5225v3 [astro-ph.CO]. doi:10.1088/0067-0049/208/2/20

[3] Planck Collaboration. (2014). Planck 2013 Results. XXIII. Isotropy and Statistics of the CMB. Astronomy & Astrophysics, A23: 1–48. https://doi.org/10.1015/0004-6361/201321534

[4] Keating, Brian. *Losing the Nobel Prize: A Story of Cosmology, Ambition, and the Perils of Science's Highest Honor.* New York: W.W. Norton & Co., 2018.

[5] Guth, Alan. *The Inflationary Universe: The Quest for a New Theory of Cosmic Origins*. New York: Basic Books, 1997.

[6] Penrose, R. (1965). Gravitational Collapse and Space-time Singularities. Phys. Rev. Lett., 14: 57.

[7] Hawking, S. and Penrose, R. (1970). The Singularities of Gravitational Collapse and Cosmology. Proc. Roy. Soc. Lond. A, 314: 529–548.

[8] Tatum, E.T., Seshavatharam, U.V.S. and Lakshminarayana, S. (2015). The Basics of Flat Space Cosmology. International Journal of Astronomy and Astrophysics, 5: 116–124. http://doi.org/10.4236/ijaa.2015.52015

[9] Tatum, E.T., Seshavatharam, U.V.S. and Lakshminarayana, S. (2015). Thermal Radiation Redshift in Flat Space Cosmology. Journal of Applied Physical Science International, 4(1): 18–26.

[10] Tatum, E.T., Seshavatharam, U.V.S. and Lakshminarayana, S. (2015). Flat Space Cosmology as an Alternative to LCDM Cosmology. Frontiers of Astronomy, Astrophysics and Cosmology, 1(2): 98–104. http://pubs.sciepub.com/faac/1/2/3 doi:10.12691/faac-1-2-3

[11] Tatum, E.T., Seshavatharam, U.V.S. and Lakshminarayana, S. (2015). Flat Space Cosmology as a Mathematical Model of Quantum Gravity or Quantum Cosmology. International Journal of Astronomy and Astrophysics, 5: 133–140. http://doi.org/10.4236/ijaa.2015.53017

[12] Andreoli, C. et al. (2020). Hubble Makes Surprising Find in Early Universe. https://www.esa.int/ESA_Multimedia/Images/2020/06/Hubble_Makes_Surprising_Find_in_Early_Universe

[13] Hashimoto, T., et al. (2018). The Onset of Star Formation 250 Million Years After the Big Bang. arXiv.1805.05966v1 [astro-ph.GA] 15 May 2018.

[14] Natarajan, P. (2018). The First Monster Black Holes. Scientific American, 318, 2, 24–29.

[15] Tatum, E.T. and Seshavatharam, U.V.S. (2018). Temperature Scaling in Flat Space Cosmology in Comparison to Standard Cosmology. Journal of Modern Physics, 9: 1404–1414. https://doi.org/10.4236/jmp.2018.97085

[16] Bowman, J.D. (2018). An Absorption Profile Centered at 78 Megahertz in the Sky-Averaged Spectrum. Nature, 555: 67–70. doi:10.1038/nature25792

[17] de Bernardis, P., et al. (2000). A Flat Universe from High-Resolution Maps of the Cosmic Microwave Background Radiation. arXiv:astro-ph/0004404v1. https://doi.org/10.1038/35010035

[18] Bucher, M. (2015). Physics of the Cosmic Microwave Background Anisotropy. arXiv:1501.04288v1

[19] Wright, E.L., et al. (1996). Angular Power Spectrum of the Cosmic Microwave Background Anisotropy Seen by the COBE DMR. Astrophysical Journal, 464: L21–L24. https://doi.org/10.1086/310073

[20] Tatum, E.T. (2018). Calculating Radiation Temperature Anisotropy in Flat Space Cosmology. Journal of Modern Physics, 9:1946–1953. https://doi.org/10.4236/jmp.2018.910123

[21] Aghanim, N., et al. (2018). Planck 2018 Results VI. Cosmological Parameters. http://arXiv:1807.06209v1

[22] Macaulay, E., et al. (2018). First Cosmological Results Using Type Ia Supernovae from the Dark Energy Survey: Measurement of the Hubble Constant. arXiv:1811.02376v1

[23] Penrose, Roger. *Fashion, Faith, and Fantasy in the New Physics of the Universe*. Princeton: Princeton University Press, 2016.

[24] Tatum, E.T. and Seshavatharam, U.V.S. (2018). Clues to the Fundamental Nature of Gravity, Dark Energy and Dark Matter. Journal of Modern Physics, 9: 1469–1483. https://doi.org/10.4236/jmp.2018.98091

[25] Verlinde, E. (2011). On the Origin of Gravity and the Laws of Newton. Journal of High Energy Physics, 4: 29–55. arXiv:1001.0785v1 [hep-th]. doi:10.1007/JHEP04(2011)029

[26] Verlinde, E. (2016). Emergent Gravity and the Dark Universe. arXiv: 1611.02269v2 [hep-th].

[27] Brouwer, M.M., et al. (2016). First Test of Verlinde's Theory of Emergent Gravity Using Weak Gravitational Lensing Measurements. Monthly Notices of the Royal Astronomical Society, 000: 1–14. arXiv:1612.03034v2 [astro-ph.CO].

[28] Weinberg, S. (1989). The Cosmological Constant Problem. Rev. Mod. Phys., 61: 1–23.

[29] Carroll, S. (2001). The Cosmological Constant. Living Rev. Relativity, 4: 5–56.

[30] Tutusaus, I., et al. (2017). Is Cosmic Acceleration Proven by Local Cosmological Probes? Astronomy & Astrophysics, 602_A73. arXiv:1706.05036v1 [astro-ph.CO].

[31] Dam, L.H., et al. (2017). Apparent Cosmic Acceleration from Type Ia Supernovae. Mon. Not. Roy. Astron. Soc. arXiv:1706.07236v2 [astro-ph.CO]

[32] Nielsen, J.T., et al. (2015). Marginal Evidence for Cosmic Acceleration from Type Ia Supernovae. Scientific Reports, 6: Article number 35596. doi:10.1038/srep35596. arXiv:1506.01354v1.

[33] Jun-Jie Wei, et al. (2015). A Comparative Analysis of the Supernova Legacy Survey Sample with ΛCDM and the R_h = ct Universe. Astronomical Journal, 149: 102 (11pp).

[34] Melia, F. (2012). Fitting the Union 2.1 SN Sample with the Rh = ct Universe. Astronomical Journal, 144: arXiv:1206.6289 [astro-ph.CO]

[35] Tatum, E.T. and Seshavatharam, U.V.S. (2018). How a Realistic Linear R_h = ct Model of Cosmology Could Present the Illusion of Late Cosmic Acceleration. Journal of Modern Physics, 9: 1397–1403. https://doi.org/10.4236/jmp.2018.97084

[36] Perlmutter, S., et al. (1999). The Supernova Cosmology Project, Measurements of Omega and Lambda from 42 High-Redshift Supernovae. Astrophysical Journal, 517: 565–586. [DOI], [astro-ph/9812133].

[37] Schmidt, B. et al. (1998). The High-Z Supernova Search: Measuring Cosmic Deceleration and Global Curvature of the Universe Using Type Ia Supernovae. Astrophysical Journal, 507: 46–63.

[38] Riess, A.G., et al. (1998). Observational Evidence from Supernovae for an Accelerating Universe and a Cosmological Constant. Astronomical Journal, 116(3): 1009–38.

[39] Guth, A.H. (1981). Inflationary Universe: A Possible Solution to the Horizon and Flatness Problems. Phys. Rev. D, 23: 347.

[40] Albrecht, A. and Steinhardt, P.J. (1982). Cosmology for Grand Unified Theories with Radiatively-Induced Symmetry Breaking. Physical Review Letters, 48: 1220–3.

[41] Linde, A.D. (1982). A New Inflationary Universe Scenario: A Possible Solution of the Horizon, Flatness, Homogeneity, Isotropy, and Primordial Monopole Problems. Physics Letters, 108B: 389–92.

[42] Steinhardt, P.J. (2011). The Inflation Debate: Is the Theory at the Heart of Modern Cosmology Deeply Flawed? Scientific American, 304(4): 18–25.

[43] Sapar, A. (2019). A Perpetually Mass-Generating Planckian Universe. Proceedings of the Estonian Academy of Sciences, 68(1): 1–12. https://doi.org/10.3176/proc.2019.1.01

CHAPTER 3

Temperature Scaling in Flat Space Cosmology in Comparison to Standard Cosmology

Abstract: Temperature scaling is redefined in this chapter in terms of a new 'Universal Temperature' T_u scale according to $T_u = T^2$, where T^2 is in K^2. This rescaling puts FSC cosmic temperature, time, total matter mass, and Hubble radius on the same scale, covering roughly 60.63 logs of 10 from the Planck epoch scale to the present scale. This chapter focuses on the relatively subtle temperature curve differences between the FSC model and the standard model of cosmology. These changes become more pronounced in the early universe. Recent observational studies of the early universe have surprised standard model proponents as to how soon the first quasars and galaxies occurred following the Big Bang. By the relative time length (14.6 billion years as opposed to 13.8 billion years in the standard model) and slope of its temperature/time curve, FSC allows for considerably more time between the formation of the first stars at cosmic dawn and the formation of the first quasars and galaxies.*

Keywords: Cosmology Theory; General Relativity; CMB Temperature; Cosmic Flatness; Black Holes; Universal Temperature; Holographic Cosmology

INTRODUCTION AND BACKGROUND

Flat Space Cosmology (FSC) was developed as a heuristic of the Hawking-Penrose idea of treating the universe expanding from a singularity state

---

*Originally published on June 8, 2018 in Journal of Modern Physics (see Appendix refs).

as being equivalent to a time-reversed gigantic black hole (*i.e.,* one which smoothly expands from a singularity as opposed to smoothly collapsing to a singularity). Penrose had started the development with his initial paper on gravitational collapse and space-time singularities [1]. Hawking's doctoral thesis [2] advanced the idea by implying the validity, within general relativity, of the black hole time-reversal idea. And finally, FSC has shown very clearly that the appropriate scaling black hole equations are remarkably accurate in modeling our expanding universe [3][4][5].

As shown in Chapter eleven, we have integrated the FSC assumptions into the Friedmann equations incorporating a cosmological term for vacuum (dark) energy. One of the results of this is that the following relation holds true in FSC:

$$\frac{3H^2c^2}{8\pi G} \cong \frac{\Lambda c^4}{8\pi G} \tag{1}$$

This equation is a consequence of modeling a spatially-flat universe from its inception. It should be remembered that, *to date, there is no observational proof that our expanding universe has been anything other than spatially-flat.* At least as far back in time as the CMB radiation release event ('recombination epoch'), our universe appears to be spatially-flat [6]. The extreme flatness of our universe presents a 'cosmic flatness problem,' first elucidated by physicist Robert Dicke in the late 1960's [7][8]. In fact, theories of cosmic inflation [9][10][11][12] were invented primarily to address this problem.

The FSC model, in sharp contrast to inflationary theory, tackles the flatness problem in an entirely different way. Perpetual spatial flatness on a *global* scale (*i.e.,* for a cosmic model as a whole) is not forbidden in a general relativity model of a finite, *continually* mass-generating, expanding universe, such as FSC. Hoyle's 'steady state' continually mass-generating model failed primarily because he didn't factor in expansion from a hotter and denser early universe. The discovery of the Cosmic Microwave Background disproved his theory. The FSC model has an entirely different set of assumptions which allow not only for a hotter and denser early universe, but also for a curvature term k in its Friedmann equations to be perpetually set to zero. Thus, equation (1) above is a direct consequence of the FSC assumptions.

In so many words, equation (1) says that, in the FSC general relativity model, the magnitude of the positive (*i.e.*, gravitationally attracting) matter energy density is always equal in absolute magnitude to the negative (dark) energy density. This is fully understandable in terms of general relativity because *a globally spatially flat universe, by definition, must have a net zero energy density. If the case were otherwise, the greater energy density term would dictate a global spatial curvature,* either positive or negative, depending upon the sign of the dominating term.

In standard cosmology, there is tension between the observations of extreme global spatial flatness and their assertion of dark energy dominance. This makes no sense. It can only be one or the other, but not both at the same time. It should be remembered that cosmic acceleration is still not proven, despite the fact that dark energy is now known to exist [13][14][15]. Realistic $R_h = ct$ universe models, such as FSC, are now the chief competition for standard cosmology. These recent statistical studies [16][17][18][19][20] of the Supernova Cosmology Project data clearly demonstrate that *cosmic acceleration is not yet proven to the exclusion of cosmic coasting models like* $R_h = ct$. FSC incorporates dark energy in perpetual perfect balance with the effect of attractive gravity on a global scale. FSC accomplishes this by

$$\Lambda \cong \frac{3H^2}{c^2} \tag{2}$$

This is the only way that equation (1) holds true. Since H_t is a changing parameter over the great span of cosmic time, *this makes FSC a dynamic dark energy model of the wCDM type, with the equation of state in FSC always defined as* $w = -1.0$. This is in keeping with the quantum field theory stipulation that the zero-state vacuum energy density must always be equal in magnitude to the vacuum pressure (*i.e.*, $\rho = p$). See Chapter eleven for details.

The purpose of the present chapter is to show how the temperature vs time curve in FSC differs slightly from that of the standard cosmology model, and to explain why this may be important in terms of observational studies of the early universe.

NEW TEMPERATURE SCALING IN FLAT SPACE COSMOLOGY

Using the thermodynamic formulae of FSC assumption 4, and incorporating the Schwarzschild formula and Hubble parameter definition of FSC assumptions 2 and 3, respectively, **Figure 1** can be created for our expanding cosmic model.

This graph shows the FSC inter-relationships of the cosmic temperature in the Kelvin scale, cosmic time, total cosmic mass of gravitational matter, and the cosmic Hubble radius. Notice that, while the Kelvin scale temperature decreases by approximately 30.315 logs of 10 from the Planck scale to the scale of the current observable universe, the parameters represented on the other three axes scale by approximately 60.63 logs of 10. This is in keeping with the following formulae which are derived from the FSC assumptions:

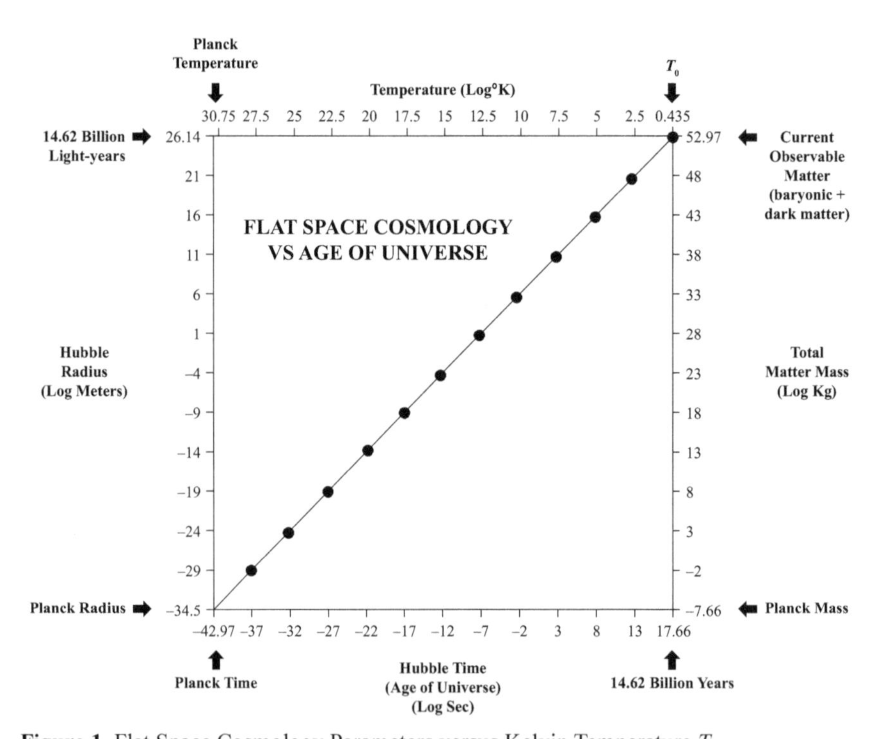

Figure 1 Flat Space Cosmology Parameters versus Kelvin Temperature *T.*

$$T^2R = 1.027246639815497 \times 10^{27}\,\text{K}^2.\text{m} \quad (3)$$

$$T^2t_s = 3.426525959553982 \times 10^{18}\,\text{K}^2.\text{s} \quad (4)$$

$$T^2t_{ys} = 1.085781647371578 \times 10^{11}\,\text{K}^2.\text{yr (sidereal years)} \quad (5)$$

$$T^2t_{yj} = 1.085781646054745 \times 10^{11}\,\text{K}^2.\text{yr (Julian years)} \quad (6)$$

In defining a new cosmological temperature scale ('Universal Temperature' T_u), by $T_u = T^2$, the Kelvin temperature scale can be converted to a finer scale which also scales by roughly 60.63 logs of 10. Therefore, the T^2R and T^2t formulae above assume the new form of T_uR and T_ut of the same numerical values. Graphs of T_u as a function of time, radius or total matter mass then become perfectly symmetrical about the x = y (*i.e.*, $T_u = t$, $T_u = R$, $T_u = M$) axis, with the vertical and horizontal axes acting as asymptotes. These graphs are shown in **Figure 2**, **Figure 3** and **Figure 4**, respectively.

Thus, the temperature/time symmetry of the FSC cosmic model can be better demonstrated. And, by incorporating the above new FSC definition of 'Universal Temperature' (*i.e.*, $T_u = T^2$), a new FSC log graph can be presented in **Figure 5**.

While there is complete one-to-one correspondence between the new T_u scale and the T (Kelvin) scale, the fineness and symmetry of scaling with T_u is considered to be an attractive feature of this approach. And yet, no information is lost by scaling the model in this way. This new temperature scaling approach opens the way for understanding FSC in terms of the holographic principle (*i.e.*, 'holographic cosmology').

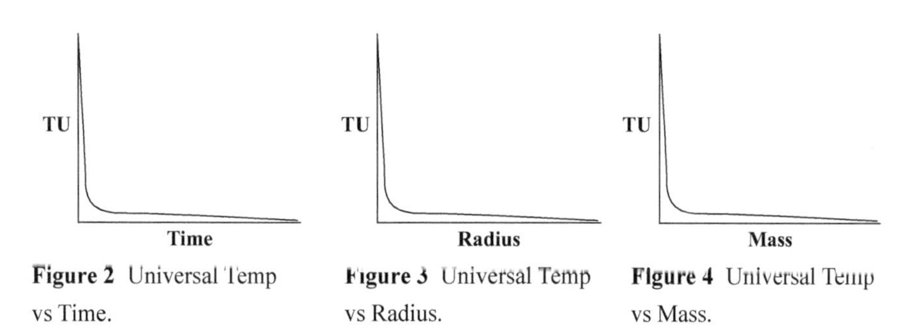

Figure 2 Universal Temp vs Time.

Figure 3 Universal Temp vs Radius.

Figure 4 Universal Temp vs Mass.

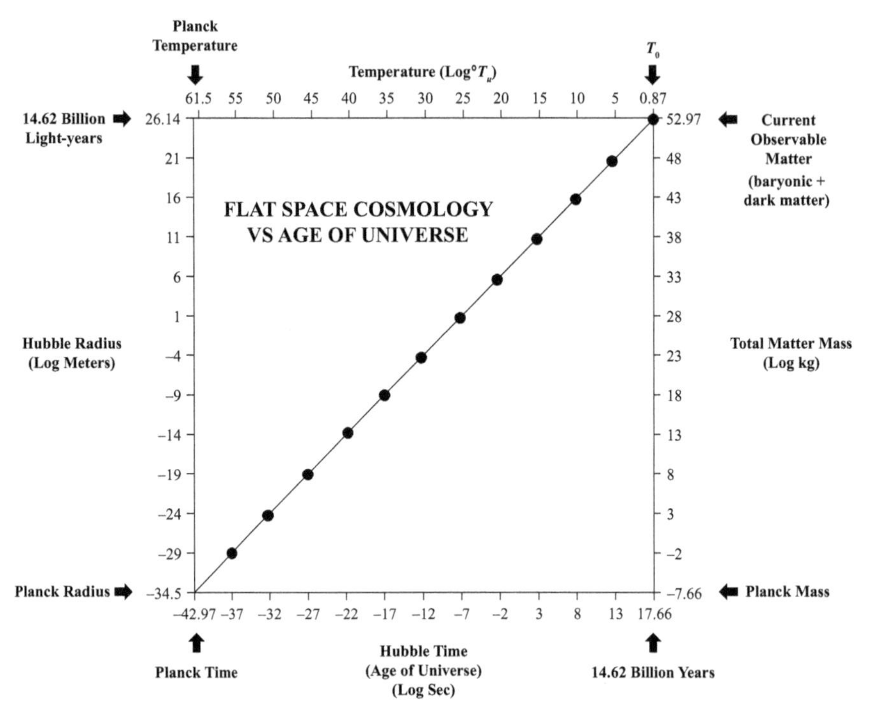

Figure 5 Flat Space Cosmology Parameters versus Universal Temperature T_U.

TEMPERATURE SCALING IN THE FSC VS STANDARD MODELS

As first reported in our thermal radiation redshift paper [4], redshift z in FSC is related to cosmic temperature by

$$z \cong \left(\frac{T_t^2}{T_o^2} - 1 \right)^{1/2} \tag{7}$$

In standard cosmology, the following formula is used

$$T_{CMB} \cong 2.725(1 + z) \tag{8}$$

wherein T_{CMB} represents the cosmic microwave background radiation temperature [21]. These temperature vs redshift formulae provide for temperature curves which reflect subtle but important differences between FSC and standard cosmology in the very early universe. However, they are negligible at the low non-zero z values of the later universe.

The temperature differences between these two models may be important with respect to the timing of the 'cosmic dawn' epoch emerging from the 'dark age' epoch. **Figure 6** shows how these models differ with respect to this transition period as a function of cosmic years after the Planck scale epoch ('Big Bang').

The upper gray line represents the standard model CMB radiation temperature and the lower gray line represents the FSC model CMB radiation temperature. The dashed gray curve represents the observed spin temperature [22], which is also presumed to be the primordial atomic hydrogen gas temperature during the earliest period of star formation. Hence the name 'cosmic dawn.'

The precipitous hydrogen gas temperature dip is believed by standard model proponents to have been triggered by temperature-induced interaction of the hydrogen gas with (presumed) colder dark matter. It is proposed that the beginning of this event can be measured by 21 cm radiation redshifted at about $z = 20$. This period of cosmic dawn appears to have ended at approximately $z = 15$. What is notable about the FSC temperature curve in comparison to the standard model curve is that *the FSC model*

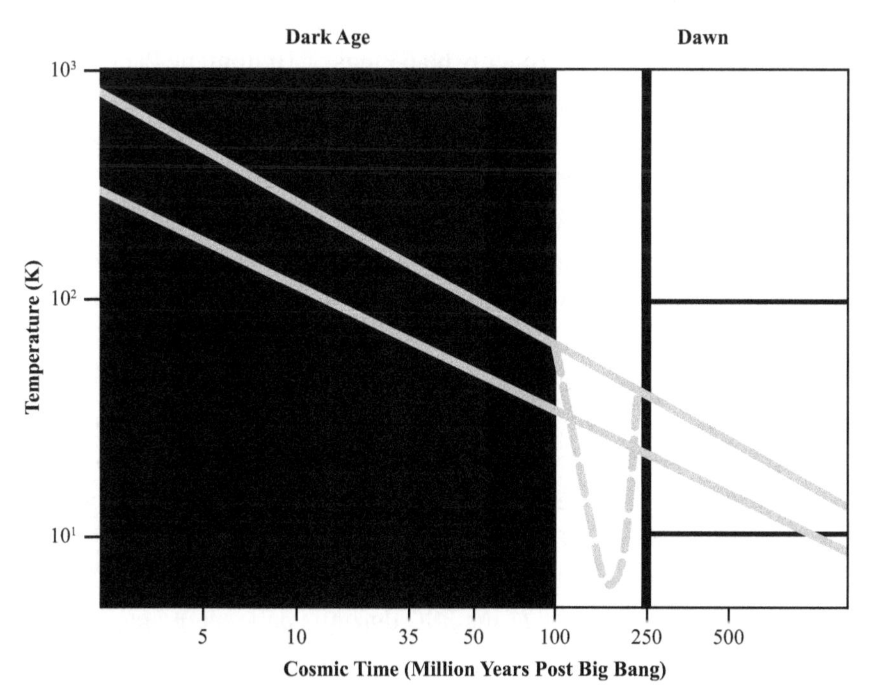

Figure 6 Temperature vs Time in Standard (upper) and FSC (lower) Models.

cosmic dawn (i.e., 20 < z < 15) would have begun as early as about 35 million years after the Planck scale epoch, as opposed to the proposed standard model cosmic dawn beginning at about 100 million years.

If one wishes to compare both models to an even earlier event in cosmic time, consider the 'recombination epoch (photon decoupling)' event which emitted the CMB radiation and is supposed to have begun at about the cosmic temperature of 3000 K. The standard model suggests this event to have occurred at approximately 379,000 years after the Big Bang, while the FSC model, by equation (5), indicates this event to have occurred approximately 12,064 years after the Big Bang. Here, of course, one is assuming an operational definition that the Big Bang occurred at approximately the Planck scale epoch. The beginning of the recombination epoch is earlier than can be included in **Figure 6**.

SUMMARY AND CONCLUSIONS

Temperature scaling is redefined in this chapter in terms of a new 'Universal Temperature' T_u scale according to $T_u = T^2$, where T^2 is in K^2. This rescaling puts FSC cosmic temperature, time, total matter mass, and Hubble radius on the same scale, covering roughly 60.63 logs of 10 from the Planck scale epoch to the present epoch.

Standard model proponents have great difficulty explaining what appears to be an impossibly short time interval between the formation of the first stars at cosmic dawn and the formation of the first quasars and galaxies. Since new cosmic epochs begin when the falling cosmic temperature reaches some threshold value, the most likely explanation for these recent observational surprises is that the standard model temperature vs cosmic time curve is slightly higher than the actual curve. By the relative time length (14.6 billion years as opposed to 13.8 billion years in the standard model) and slope of its temperature/time curve, FSC allows for considerably more time between the formation of the first stars at cosmic dawn and the formation of the first quasars and galaxies.

Further observational studies of the early universe are expected to ultimately prove the superiority of the FSC dynamic dark energy model with respect to the standard model [23].

REFERENCES

[1] Penrose, R. (1965). Gravitational Collapse and Space-time Singularities. Phys. Rev. Lett., 14: 57.

[2] Hawking, S. and Penrose, R. (1970). The Singularities of Gravitational Collapse and Cosmology. Proc. Roy. Soc. Lond. A, 314: 529–548.

[3] Tatum, E.T., Seshavatharam, U.V.S. and Lakshminarayana, S. (2015). The Basics of Flat Space Cosmology. International Journal of Astronomy and Astrophysics, 5: 116–124. http://doi.org/10.4236/ijaa.2015.52015

[4] Tatum, E.T., Seshavatharam, U.V.S. and Lakshminarayana, S. (2015). Thermal Radiation Redshift in Flat Space Cosmology. Journal of Applied Physical Science International, 4(1): 18–26.

[5] Tatum E.T., Seshavatharam, U.V.S. and Lakshminarayana, S. (2015). Flat Space Cosmology as an Alternative to LCDM Cosmology. Frontiers of Astronomy, Astrophysics and Cosmology, 1(2): 98–104. http://pubs.sciepub.com/faac/1/2/3

[6] Planck Collaboration XIII. (2016). Cosmological Parameters. Astronomy & Astrophysics, 594, A13. doi:10.1051/0004-6361/201525830. http://arxiv.org/abs/1502.01589

[7] Dicke, R.H. (1970). Gravitation and the Universe. American Philosophical Society.

[8] Guth, Alan. *The Inflationary Universe: The Quest for a New Theory of Cosmic Origins*. New York: Basic Books, 1997.

[9] Guth, A.H. (1981). Inflationary Universe: A Possible Solution to the Horizon and Flatness Problems. Phys. Rev. D, 23: 347.

[10] Albrecht, A. and Steinhardt, P.J. (1982). Cosmology for Grand Unified Theories with Radiatively-Induced Symmetry Breaking. Physical Review Letters, 48: 1220–3.

[11] Linde, A.D. (1982). A New Inflationary Universe Scenario: A Possible Solution of the Horizon, Flatness, Homogeneity, Isotropy, and Primordial Monopole Problems. Physics Letters, 108B: 389–92.

[12] Linde, Andre. *Inflation and Quantum Cosmology*. Boston: Academic Press, 1990.

[13] Perlmutter, S., et al. (1999). The Supernova Cosmology Project, Measurements of Omega and Lambda from 42 High-Redshift Supernovae. Astrophysical Journal, 517: 565–586. [DOI], [astro-ph/9812133].

[14] Schmidt, B., et al. (1998). The High-Z Supernova Search: Measuring Cosmic Deceleration and Global Curvature of the Universe Using Type Ia Supernovae. Astrophysical Journal, 507: 46–63.

[15] Riess, A.G., et al. (1998). Observational Evidence from Supernovae for an Accelerating Universe and a Cosmological Constant. Astronomical Journal, 116(3): 1009–38.

[16] Tutusaus, I., et al. (2017). Is Cosmic Acceleration Proven by Local Cosmological Probes? Astronomy & Astrophysics, 602_A73.arXiv:1706.05036v1 [astro-ph.CO].

[17] Dam, L.H., et al. (2017). Apparent Cosmic Acceleration from Type Ia Supernovae. Mon. Not. Roy. Astron. Soc. arXiv:1706.07236v2 [astro-ph.CO]

[18] Nielsen, J.T., et al. (2015). Marginal Evidence for Cosmic Acceleration from Type Ia Supernovae. doi:10.1038/srep35596. arXiv:1506.01354v1.

[19] Jun-Jie Wei, et al. (2015). A Comparative Analysis of the Supernova Legacy Survey Sample with ΛCDM and the R_h=ct Universe. *Astronomical Journal*, 149: 102–112.

[20] Melia, F. (2012). Fitting the Union 2.1 SN Sample with the Rh=ct Universe. Astronomical Journal, 144. arXiv:1206.6289 [astro-ph.CO].

[21] Barkana, R. (2018). Cosmic Dawn as a Dark Matter Detector. arXiv:1803.06698v1 [astro-ph.CO].

[22] Bowman, J.D. (2018). An Absorption Profile Centered at 78 Megahertz in the Sky-Averaged Spectrum. Nature, 555: 67–70. doi:10.1038/nature25792

[23] Zhao, G., et al. (2017). Dynamical Dark Energy in Light of the Latest Observations. arXiv:1701.08165v2 [astro-ph.CO].

CHAPTER 4

How a Realistic Linear $R_h = ct$ Model of Cosmology May Present the Illusion of Late Cosmic Acceleration

Abstract: Realistic FLRW cosmic coasting models which contain matter now appear to be a reasonable alternative in explaining the accumulated Supernova Cosmology Project data since 1998. In sharp contrast to the unrealistic original classic Milne universe, which was entirely devoid of matter, these modified Milne-type models containing matter, often referred to as realistic linear $R_h = ct$ models, have rapidly become the primary competition with standard cosmology. This chapter compares the expected relative luminosity distances and relative angular diameter distances for given magnitudes of redshift within these two competing models. A simple ratio formula is derived, which explains how *expected* luminosity distances and angular diameter distances for given magnitudes of redshift within a realistic Milne-type cosmic expansion could create the illusion (for standard model proponents) of cosmic acceleration where none exists.*

Keywords: Dark Energy; Cosmology Theory; Cosmic Coasting; Cosmic Flatness; Type Ia Supernovae; CMB; Flat Space Cosmology; Milne Universe

---

*Originally published on June 8, 2018 in Journal of Modern Physics (see Appendix refs).

INTRODUCTION AND BACKGROUND

While final observations of the Dark Energy Survey (DES) are eagerly awaited, there is a vigorous debate in the scientific community as to whether cosmic acceleration is a reality or an illusion. Some very recent papers [1][2][3][4][5] present compelling statistical analysis of the accumulated data which clearly shows that *cosmic acceleration is not yet proven*. In particular, it appears that constant velocity cosmic expansion (cosmic 'coasting') could produce very similar observations [6][7][8][9] [10][11][12][13][14].

Until very recently, it was believed that *only* a classic Milne ('empty universe') model could produce cosmic coasting. Unfortunately, Milne's original model [15] was unrealistic, being entirely devoid of matter throughout its expansion history. Initially after the 1998 Type Ia supernovae discovery of dark energy [16][17][18], there appeared to be no realistic cosmic models containing matter which could compete with the current widely-accepted standard Friedmann-Lemaitre-Robertson-Walker (FLRW) Lambda Cold Dark Matter (ΛCDM) model (hereafter referred to as the 'standard model'). Fortunately, within the last few years, a great deal of progress has been made in developing realistic FLRW cosmic coasting models which contain matter. *These $R_h = ct$ models have rapidly become the primary competition with the standard model.*

As detailed in Dam, *et al.* (2017), the luminosity distance (d_L) redshift relation in the standard model is given exactly by

$$
\begin{aligned}
d_L &= \frac{(1+z)c}{H_0\sqrt{|\Omega_{k0}|}}\left(\operatorname{sinn}\int_{1/(1+z)}^{1}\sqrt{|\Omega_{k0}|}\frac{dy}{H(y)}\right),\\
H(y) &\equiv \sqrt{\Omega_{R0}+\Omega_{M0}y+\Omega_{k0}y^2+\Omega_{\Lambda 0}y^4}, \qquad (1)\\
\operatorname{sinn}(x) &\equiv \begin{cases}\sinh(x), & \Omega_{k0}>0\\ x, & \Omega_{k0}=0,\\ \sin(x), & \Omega_{k0}<0\end{cases}
\end{aligned}
$$

In the standard model this equation simplifies to $d_L = (1+z)(c/H_0)$ for the extended period of cosmic expansion history in which space is essentially flat. Judging from recent Cosmic Microwave Background (CMB) observations and current estimates of the Omega cosmic density parameter [19], this period of flat space cosmic expansion appears to have been present since at least the recombination epoch.

As also detailed in Dam, *et al.* (2017), luminosity distance in the realistic Milne-type model with linear expansion (*i.e.*, with $a(t)$ proportional to t) can be compared to luminosity distance in relation (1). Notably, *the realistic Milne-type model can have any matter content, so long as the luminosity distance is exactly*

$$d_L = z\left(1+\frac{z}{2}\right)\left(\frac{c}{H_0}\right) \tag{2}$$

In the following presentation, the above two mathematical expressions of luminosity distance observed within these two distinctly different cosmological models are compared and the important implications discussed.

RATIO OF STANDARD MODEL LUMINOSITY DISTANCE TO REALISTIC MILNE-TYPE MODEL LUMINOSITY DISTANCE

To see how the luminosity distances within both models compare to one another, they can be expressed as a ratio in the redshift terms z and s, respectively. These redshift terms are related by $s = z + 1$.

Based upon relations (1) and (2), the ratio of the standard model luminosity distance to the realistic Milne-type model luminosity distance, expressed as a function of redshift term z is given by

$$\frac{\text{LCDM } d_L}{\text{Milne-type } d_L} = (1+z)\Big/ z\left(1+\left(\frac{z}{2}\right)\right) \tag{3}$$

With substitution of redshift term s for redshift term z, this ratio of the standard model luminosity distance to the realistic Milne-type model luminosity distance is also given by

$$\frac{\text{LCDM } d_L}{\text{Milne-type } d_L} = \frac{2s}{(s^2 - 1)} \tag{4}$$

Using relation (4), one can easily see that the luminosity distances in the two models are equal *only* when quadratic equation $s^2 - 2s - 1 = 0$ is solved for its positive value of $s = (\sqrt{2} + 1)$. Thus, the luminosity distances in the two models are also equal when $z = \sqrt{2}$. At this magnitude of redshift is the luminosity distance point of intersection for these models.

DISCUSSION

Since the recent publications of FLRW linear coasting cosmology models, it has become especially important to compare such modified Milne-type models to the standard model. In this paper, particular attention is paid to how luminosity distances and angular diameter distances compare between these two models. One can easily see from relations (3) and (4) that, *beyond the redshift value at the point of intersection, the linear model luminosity distance value is always greater than the luminosity distance expected within the standard model. It is also evident that the difference between luminosity distances within these two models becomes ever larger as cosmological redshifts increase.* The relative luminosity distances (in arbitrary units) are shown in **Figure 1**.

In FLRW models, the following relations (5) and (6) apply for luminosity distance $d_L(z)$ and angular diameter distance $d_A(z)$, respectively:

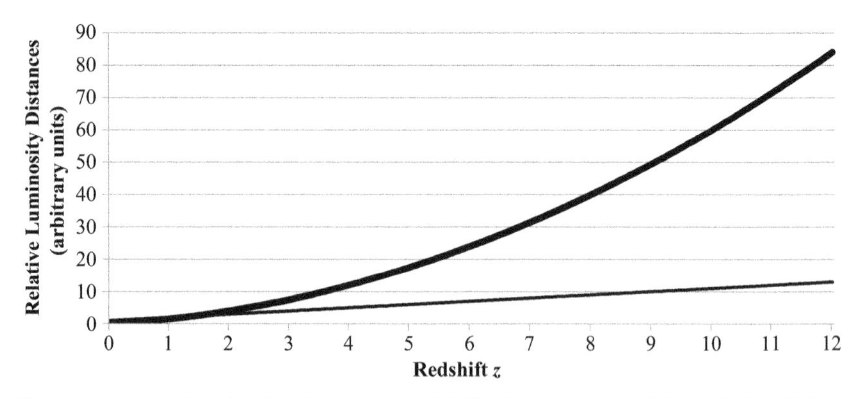

Figure 1 Relative luminosity distances vs redshift z for standard (thin) and Milne (thick) models.

Luminosity distance:

$$d_L(z) = (1+z)d_M(z) \tag{5}$$

Angular diameter distance:

$$d_A(z) = \frac{d_M(z)}{(1+z)} \tag{6}$$

In these relations, $d_M(z)$ is the transverse co-moving distance which, during the extremely flat (*i.e.*, $\Omega_K = 0$) cosmic expansion period extending from at least the recombination epoch to the present, equates to the co-moving distance [*i.e.*, $d_M(z) = d_C(z)$].

From relations (5) and (6) it is obvious that:

$$\frac{d_L(z)}{d_A(z)} = (1+z)^2 = s^2 \tag{7}$$

Therefore, *in a comparison of any two FLRW models, the relative angular diameter distances pertaining to any given cosmological redshift maintain exactly the same relationship (i.e., ratio) as do the relative luminosity distances.* This becomes the basis for **Figure 2**:

For any FLRW model, linear or otherwise, to compete with the standard model, three major considerations must be addressed:

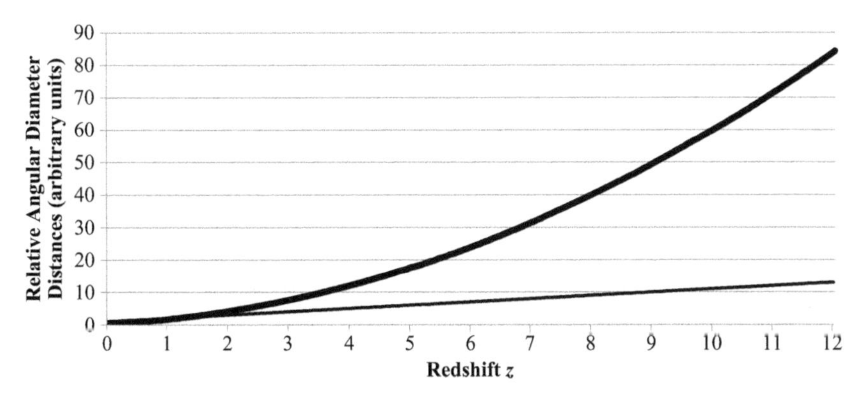

Figure 2 Relative angular diameter distances vs redshift z for standard (thin) and Milne (thick) models.

1) What is the capability of expected future observations to discriminate between the two models?
2) Given that the Type Ia supernovae data are just one piece of the cosmological data at present, does the competing model fall within other existing observational constraints, including those of baryonic acoustic oscillations (BAO) and the cosmic microwave background (CMB)?
3) Do the Planck constraints on Ω_m possibly fit with the competing model in question?

In realistic linear $R_h = ct$ models these considerations can be addressed as follows:

1) It may not yet be possible to propose a definitive observational study between these two competing models. Nevertheless, the *theoretical* analysis in the present paper shows the importance of not yet being completely dogmatic about one particular model to the exclusion of all others, particularly when the nature of the non-matter energy currently known as 'dark energy' is completely mysterious to us. Perhaps, someday, we'll be left only with Occam's razor to choose which of the two models makes the most sense to us.
2) With respect to the question of observational constraints and realistic linear (*i.e.*, flat) models, Figure 5 of the following link to the Supernova Cosmology Project is offered as proof: https://dx.doi.org/10.1088/0004-637x/746/1/85 [20]. One can readily see (by the 'Flat' line intersection) that a realistic spatially-flat universe model such as FSC is an excellent fit with all such observations to date. It is obvious, judging from the intersections of the 'Flat' universe model line, that *realistic* linear flat space cosmology models would fit within the highly-constrained zone of intersection, and cannot be dismissed out of hand by the accumulated data so far. This conclusion is also strongly supported by the data analysis of Tutusaus, *et al.* (2017).
3) Given the current constraints on Ω_m presented in the Supernova Cosmology Project figure, for a realistic linear $R_h = ct$ model (which is

flat and *non-accelerating* by definition), the non-matter proportion of cosmic energy density, represented by Ω_Λ, would presumably be latent and stored within the cosmic vacuum, perhaps as virtual particles. This is speculation, of course, at the present time. Within a linear cosmic coasting model there is no *net* force acting upon the cosmic expansion. Obviously, whatever the nature of dark energy happens to be, it requires further explanation within *both* accelerating and coasting FLRW models.

CONCLUSIONS

The significance of the relative luminosity distance and relative angular diameter distance comparisons between these two competing models is paramount. If an observer of distant Type Ia supernovae expects particular luminosity distances, or angular diameter distances, corresponding to particular redshifts and, instead, sees greater-than-expected luminosity distances (*i.e.*, unexpected 'dimming' of the supernovae) or greater-than-expected angular diameter distances, this can easily be misinterpreted by a standard model proponent as indicative of cosmic acceleration. *Entirely predictable supernova luminosity distances within a realistic Milne-type universe containing matter, as opposed to a standard model universe, could thus be one possible explanation for the Type Ia supernovae observations since 1998. Obviously, cosmic acceleration would not then be required to explain these observations.*

An interesting feature of this luminosity distance and angular diameter distance comparison is the point of intersection at $s = (\sqrt{2} + 1)$ and $z = \sqrt{2}$. It is expected that the DES study will ultimately show what would appear to standard model proponents as gradually accelerating cosmic expansion beyond this point of intersection at roughly 6 billion years of cosmic age. However, when the DES study is completed, one must be very careful to still consider the possibility of a linear $R_h = ct$ cosmic expansion presenting the illusion of late cosmic acceleration, as revealed mathematically herein, if the DES observations are interpreted (possibly incorrectly) within the context of standard cosmology.

REFERENCES

[1] Tutusaus, I., et al. (2017). Is Cosmic Acceleration Proven by Local Cosmological Probes? Astronomy & Astrophysics, 602_A73. arXiv:1706.05036v1 [astro-ph.CO].

[2] Dam, L.H., et al. (2017). Apparent Cosmic Acceleration from Type Ia Supernovae. Mon. Not. Roy. Astron. Soc. arXiv:1706.07236v2 [astro-ph.CO]

[3] Nielsen, J.T., et al. (2015). Marginal Evidence for Cosmic Acceleration from Type Ia Supernovae. Scientific Reports, 6: Article number 35596. doi: 10.1038/srep35596. arXiv:1506.01354v1.

[4] Jun-Jie Wei., et al. (2015). A Comparative Analysis of the Supernova Legacy Survey Sample with ΛCDM and the R_h=ct Universe. Astronomical Journal, 149: 102 (11pp).

[5] Melia, F. (2012). Fitting the Union 2.1 SN Sample with the R_h=ct Universe. Astronomical Journal, 144: arXiv:1206.6289 [astro-ph.CO]

[6] Gehlaut, S., et al. (2002). A "Freely Coasting" Universe. arXiv:astro-ph/0209209v.2

[7] Melia, F., et al. (2012). The Rh=ct Universe. Mon. Not. Roy. Astron. Soc. 419: 2579–2586. arXiv:1109.5189 [astro-ph.CO]

[8] Melia, F., et al. (2013). Cosmic Chronometers in the Rh=ct Universe. Mon. Not. Roy. Astron. Soc., 432: 2669–2675. arXiv:1304.1802 [astro-ph.CO]

[9] Melia, F. (2015). On Recent Claims Concerning the R_h=ct Universe. Mon. Not. Roy. Astron. Soc., 446: 1191–1194. arXiv:1406.4918 [astro-ph.CO]

[10] Melia, F., et al. (2016). The Epoch of Reionization in the R_h=ct Universe. Mon. Not. Roy. Astron. Soc., 456(4): 3422–3431. arXiv:1512.02427 [astro-ph.CO]

[11] Tatum, E.T., Seshavatharam, U.V.S. and Lakshminarayana, S. (2015). The Basics of Flat Space Cosmology. International Journal of Astronomy and Astrophysics, 5: 116–124. http://doi.org/10.4236/ijaa.2015.52015

[12] Tatum, E.T., Seshavatharam, U.V.S. and Lakshminarayana, S. (2015). Thermal Radiation Redshift in Flat Space Cosmology. Journal of Applied Physical Science International, 4(1): 18–26.

[13] Tatum, E.T., Seshavatharam, U.V.S. and Lakshminarayana, S. (2015). Flat Space Cosmology as an Alternative to LCDM Cosmology. Frontiers of Astronomy, Astrophysics and Cosmology, 1(2): 98–104. http://pubs.sciepub.com/faac/1/2/3

[14] John, M.V. (2016). Realistic Coasting Cosmology from the Milne Model. Mon. Not. Roy. Astron. Soc., 000: 1–12. arXiv:1610.09885v1 [astro-ph.CO]

[15] Milne, E.A. (1933). Z. Astrophysik, 6: 1–35.

[16] Perlmutter, S., et al. (1999). The Supernova Cosmology Project, Measurements of Omega and Lambda from 42 High-Redshift Supernovae. Astrophysical Journal, 517: 565–586. [DOI], [astro-ph/9812133].

[17] Schmidt, B., et al. (1998). The High-Z Supernova Search: Measuring Cosmic Deceleration and Global Curvature of the Universe Using Type Ia Supernovae. Astrophysical Journal, 507: 46–63.

[18] Riess, A.G., et al. (1998). Observational Evidence from Supernovae for an Accelerating Universe and a Cosmological Constant. Astronomical Journal, 116(3): 1009–38.

[19] Planck Collaboration XIII (2015). Cosmological Parameters. http://arxiv.org/abs/1502.01589

[20] Suzuki, N., et al. (2011). The Hubble Space Telescope Cluster Supernovae Survey: V. Improving the Dark Energy Constraints Above Z>1 and Building an Early-Type-Hosted Supernova Sample. arXiv.org/abs/1105.3470.

CHAPTER 5

Clues to the Fundamental Nature of Gravity, Dark Energy and Dark Matter

Abstract: This chapter integrates the Flat Space Cosmology (FSC) model, including our entropy assumption, into the Friedmann equations containing a cosmological term. The Lambda Λ term within this model scales according to $3H_t^2/c^2$ and $3/R_t^2$. Use of the Bekenstein-Hawking definition of closed gravitational system total entropy S provides for FSC cosmic parameter definitions in terms of $\sqrt{S}$. Cosmic time, radius, total matter mass-energy and vacuum energy in this model scale in exactly the same way as $\sqrt{S}$. This analysis opens the way for thinking of gravity, dark energy and dark matter as being somehow deeply connected with total cosmic entropy. The recent theoretical work of Roger Penrose and Erik Verlinde is discussed in this context. The results of this FSC model analysis dovetail nicely with Verlinde's work suggesting gravity as being fundamentally an emergent property of cosmic entropy. This emergent-property-of-entropy theory of gravity, if true, could also suggest that gravitational inertia, dark matter and dark energy might simply be manifestations of cosmic entropy. If such were the case, these entities might be difficult to define by the rules of quantum physics.*

Keywords: Cosmology Theory; Dark Energy; Dark Matter; Cosmic Entropy; Entropic Arrow of Time; Cosmic Inflation; Black Holes; Cosmological Constant Problem; Emergent Gravity; Mach's Principle

---

*Originally published on July 12, 2018 in Journal of Modern Physics (see Appendix refs).

INTRODUCTION AND BACKGROUND

Flat Space Cosmology (FSC) is a mathematical model of universal expansion which has proven to be remarkably accurate in comparison to observations [1][2][3][4][5]. FSC was initially developed as a heuristic mathematical model of the Hawking-Penrose idea that an expanding universe arising from a singularity state can be modeled as a time-reversed giant black hole. This idea was an extension of Penrose's paper [6] on the singularities of black holes and cosmology. Hawking's doctoral thesis took the idea further by proving the validity of time-reversal in the treatment of general relativity as it concerns cosmology [7]. Finally, the FSC model completes this idea by incorporating scaling black hole equations suitable for cosmology. Thus, the proven accuracy of FSC with respect to current astronomical observations does not appear to be an accident.

FSC has recently been proven to be a general relativity model by successfully integrating the FSC assumptions into the Friedmann equations which include a cosmological term and a global curvature term k set to zero. The relevant equations are repeated in this paper for clarity. One of the results of integrating FSC into the Friedmann equations is that the following relation holds true in FSC:

$$\frac{3H^2c^2}{8\pi G} \cong \frac{\Lambda c^4}{8\pi G} \tag{1}$$

This is merely a reflection that global space-time in FSC is flat during the cosmic expansion. As stipulated by the space-time curvature rules of general relativity, a globally flat universe *must* have a net energy density of zero. Otherwise, if the positive energy density and negative energy density terms were not equal in magnitude, there would be an observable global space-time curvature representative of the greater energy density term.

The purpose of this chapter is to show how the FSC Friedmann equations evolve further from equation (1) and what they might imply with respect to the fundamental nature of gravity, dark energy and dark matter. *In particular, this chapter shows how the Bekenstein-Hawking definition of black hole entropy [8] [9] can be applied in the FSC model.*

FLAT SPACE COSMOLOGY FRIEDMANN EQUATIONS

With respect to the Friedmann equations, those incorporating a non-zero cosmological term (*i.e.*, a dark energy term) are now the most relevant since the 1998 Type Ia supernovae discoveries. Therefore, accepting Friedmann's starting assumptions of homogeneity, isotropism and an expanding cosmic system with a stress-energy tensor of a perfect fluid, we have his cosmological equation

$$\frac{\dot{a}^2 + kc^2}{a^2} \cong \frac{8\pi G\rho + \Lambda c^2}{3} \tag{2}$$

This equation is derived from the 00 component of the Einstein field equations. Since the global curvature term k is always zero in FSC, equation (2) reduces to

$$\left(\frac{\dot{a}}{a}\right)^2 \cong H^2 \cong \frac{8\pi G\rho}{3} + \frac{\Lambda c^2}{3} \tag{3}$$

With rearrangement, we have

$$\frac{3H^2}{8\pi G} - \frac{\Lambda c^2}{8\pi G} \cong \rho \tag{4}$$

This is the relevant Friedmann equation for cosmic mass density. Multiplying all terms by c^2 gives us the relevant Friedmann equation for cosmic energy density

$$\frac{3H^2c^2}{8\pi G} - \frac{\Lambda c^4}{8\pi G} \cong \rho c^2 \tag{5}$$

At this point it is crucial to remember that Friedmann's energy density derivation of Einstein's field equations for the cosmic system as a whole (*i.e.*, globally) can be interpreted in the form of additive space-time curvatures represented by the individual terms. The first term can be read as the positive energy density (*i.e.*, the positive space-time curvature) term; the second term can be read as the negative energy density (*i.e.*, the negative space-time curvature) term; and the third term can be read as the summation (*i.e., net*) energy density term for global cosmic space-time curvature. Since global space-time is treated as constantly and perfectly flat in FSC, the third term must always have a net value of zero energy density. This is entirely in keeping with the general theory of relativity, as applied to cosmology, as well as current cosmological observations of flatness (*i.e.*, critical density). Hence, in FSC

$$\frac{3H^2}{8\pi G} \cong \frac{\Lambda c^2}{8\pi G} \tag{6}$$

and

$$\frac{3H^2c^2}{8\pi G} \cong \frac{\Lambda c^4}{8\pi G} \tag{7}$$

From these respective critical mass density and energy density equations, it is obvious that the FSC model defines the Lambda term Λ by

$$\Lambda \cong \frac{3H^2}{c^2} \tag{8}$$

In FSC and other realistic linear Milne-type models, Hubble parameter H is a quantity which scales with cosmic time and is defined as

$$H \cong \frac{c}{R} \tag{9}$$

where c is the speed of light and R is the cosmic radius as defined by the Schwarzschild formula

$$R \cong \frac{2GM}{c^2} \tag{10}$$

where M represents the total matter mass of the cosmic system and G is the universal gravitational constant. Therefore, FSC equation (8) substituted by equation (9) gives

$$\Lambda \cong \frac{3}{R^2} \tag{11}$$

So, the Lambda term Λ is also a scalar quantity (*i.e.,* like the Hubble parameter, not actually a constant) over the great span of cosmic time. This indicates that *FSC is a dynamic dark energy quintessence model.*

Crucially, equation (11) allows one to compare the Lambda term Λ with total entropy for the FSC cosmic system over the span of cosmic time. Recalling the Bekenstein-Hawking derivation of black hole entropy as being directly proportional to the event horizon surface area ($4\pi R^2$), we use their formula for cosmic entropy:

$$S_t \cong \frac{\pi R_t^2}{L_p^2} \tag{12}$$

Then substituting equation (11) into equation (12) and rearranging terms:

$$\Lambda \cong \frac{3\pi}{SL_p^2} \tag{13}$$

Thus, the Lambda term Λ in FSC is inversely proportional to total cosmic entropy S at all times. Substituting equation (13) into equation (8) gives

$$S \cong \frac{\pi c^2}{H^2 L_p^2} \tag{14}$$

and

$$H \cong \frac{c}{L_p}\sqrt{\frac{\pi}{S}} \tag{15}$$

And, since the reciprocal of the Hubble parameter is the measure of cosmic time t in FSC

$$t \cong \frac{L_p}{c}\sqrt{\frac{S}{\pi}} \tag{16}$$

So cosmic time is always directly proportional to $\sqrt{S}$, with entropy S as defined by Bekenstein and Hawking. Thus, the 'entropic arrow of time' is clearly defined in the FSC model.

The dark energy density cosmological term is not only expressed as $(\Lambda c^4/8\pi G)$ in FSC Friedmann equation (7) but, by incorporating equation (13) into this term, we now have a dark energy density equation

$$\frac{\Lambda c^4}{8\pi G} \cong \frac{3c^4}{8GSL_p^2} \cong \frac{3H^2c^2}{8\pi G} \tag{17}$$

wherein any of these terms can be used interchangeably to quantify the absolute magnitude of the cosmic dark energy density at all times.

Given the above relations, simple algebraic rearrangements allow for expressions of the following FSC parameters in terms of $\sqrt{S}$:

$$\sqrt{S} = \frac{c\sqrt{\pi}}{L_p}t \tag{18}$$

Showing direct proportionality between cosmic entropy and cosmic time t.

$$\sqrt{S} = \frac{\sqrt{\pi}}{L_p}R \tag{19}$$

Showing direct proportionality between cosmic entropy and cosmic radius R.

$$\sqrt{S} = \frac{2G\sqrt{\pi}}{c^2 L_p}M \tag{20}$$

Showing direct proportionality between cosmic entropy and total cosmic matter mass M.

$$\sqrt{S} = \frac{\hbar c^5}{32\pi^{\frac{3}{2}} k_B^2 G} T_U^{-1} = \frac{\hbar c^5}{32\pi^{\frac{3}{2}} k_B^2 G} T^{-2} \tag{21}$$

Showing inverse proportionality between cosmic entropy and cosmic temperatures T_U and T.

$$\sqrt{S} = \frac{c\sqrt{\pi}}{L_p} H^{-1} \tag{22}$$

Showing inverse proportionality between cosmic entropy and Hubble parameter H.

$$\sqrt{S} = \frac{\sqrt{3\pi}}{L_p} \Lambda^{-\frac{1}{2}} \tag{23}$$

Showing inverse proportionality between cosmic entropy and cosmic Lambda.

$$\sqrt{S} = \frac{2G\sqrt{\pi}}{c^4 L_p} Mc^2 \tag{24A}$$

$$\sqrt{S} = \frac{2G\sqrt{\pi}}{c^4 L_p} (V.E) \tag{24B}$$

Showing direct proportionality between cosmic entropy and total cosmic matter mass-energy and negative vacuum energy.

$$\frac{M}{R} = \frac{c^2}{2G} \tag{25}$$

Showing the Schwarzschild relation between total cosmic matter mass M and radius R.

$$\frac{GM^2}{R^2} = \frac{c^4}{4G} \tag{26}$$

Showing an FSC Newtonian gravitational force relation based upon the Schwarzschild relation.

$$Mc^2 = \frac{c^4}{2G}R \tag{27A}$$

$$|V.E| = \frac{c^4}{2G}R \tag{27B}$$

Showing FSC energy definitions of total cosmic matter mass-energy and vacuum energy.

$$\frac{Mc^2}{2} = \frac{GM^2}{R^2}R \tag{28A}$$

$$\frac{|V.E|}{2} = \frac{GM^2}{R^2}R \tag{28B}$$

Showing FSC matter mass-energy and vacuum energy relations with FSC Newtonian gravitational work (incorporating $E = Mc^2$, of course).

$$\frac{Mc^2}{2} + \frac{V.E}{2} = \left(\frac{GM^2}{R^2}\right)R - \left(\frac{GM^2}{R^2}\right)R = 0 \tag{29}$$

Showing how conservation of energy works in the expanding FSC closed energy system. Such a spatially flat cosmic system, if it *begins* with net zero energy, must *always be* at net zero energy.

DISCUSSION

Incorporation of the FSC assumptions into the Friedmann equations containing a cosmological term provides unique insights into the possible nature of gravity, dark energy and dark matter. The cosmological term is usually expressed in the form of a negative energy density in counterbalance to the positive energy density of total matter (baryonic plus dark matter). Given the recent discovery of dark energy [10][11][12], and in the context of general relativity, dark energy is believed to represent a *systemic* negative gravitational energy within the cosmological vacuum. It seems reasonable to assume that dark energy and the negative vacuum energy represented by Friedmann's cosmological term are one and the same. The important question concerns whether dark energy is a completely new physical entity or one which we already know by another name.

Energy *within the vacuum* of a closed gravitating system has long been assumed to be an opposite sign energy in comparison to matter energy. For an excellent discussion as to why such energy should be of opposite signage, the interested reader is referred to pages 11–14 and 289–293 in Alan Guth's excellent book entitled, 'The Inflationary Universe' [13]. Gravitational systems subtract potential energy from mass bodies when aggregating them. Thus, by $E = mc^2$, all aggregating bodies lose increments of mass corresponding to their decreased gravitational potential energy. By convention, this is regarded as a loss in the positive energy of matter. However, the generalized vacuum part of any such closed system must either lose an equal amount of negative energy, or gain an equal amount of positive energy, during all such gravitational interactions, in order to obey the Law of Conservation of Energy. No net energy can be gained or lost by a gravitating closed system, whether it is expanding, contracting or fixed in radius. Moreover, as the total FSC matter mass-energy *increases* over cosmic time, an opposite sign energy (call it 'dark energy') must accumulate within the cosmic vacuum.

In this context, it is easy to understand the meaning of FSC mass density and energy density equations (6) and (7), respectively. Equality between these total matter and vacuum energy terms is *mandatory* in a closed system such as FSC. And, because these terms are of opposite signs with respect to their energy densities, *the net global energy density of a spatially-flat closed gravitating system must be perpetually zero from inception.* The FSC assumptions, by virtue of the Schwarzschild formula relationship between total matter mass M_t and radius R_t, and by virtue of the Hubble parameter definition as c/R_t, create a flat universe perpetually at the Friedmann critical energy density of $3H^2c^2/8\pi G$. By incorporating the Schwarzschild relation [equation (25)] into total matter and vacuum energy equations (27A) and (27B), one can readily see how Newtonian gravitational work (now slightly modified by incorporating $E = mc^2$, of course) can be expressed in equations (28A) and (28B). Incorporating the correct negative energy signage of vacuum energy (*i.e.,* dark energy) into equation (29) shows how a closed net zero energy (*i.e.,* flat) gravitating universe could evolve from a net zero energy quantum fluctuation event.

In sharp contrast to FSC, standard inflationary cosmology has an entirely different explanation for cosmological flatness in universal observations going all the way back to the *very* early universe [the Cosmic Microwave Background (CMB) radiation was released *before* 3 one-hundred-thousandths (0.0000277) of the current age of the universe]. Standard cosmology maintains that a quantum fluctuation event within a zero energy pre-Big Bang state may have initiated the universal expansion. It also maintains that gravity was the first of four fundamental forces to 'freeze out' following an exceedingly brief exponential inflationary phase. Standard model cosmologists believe our current universe to contain an extremely small *net* negative energy. In other words, they believe in cosmic acceleration (as opposed to constant velocity light speed expansion), despite these current observations of extreme flatness. However, *if our universe began from a zero-energy state and now has a non-zero energy density, however small, this would appear to violate conservation of energy!* Furthermore, one must ask what kind of force drove the inflationary ('inflaton') field if gravity did not already exist at the inception of the universe. Cosmic inflation energy appears to be suspiciously like early cosmic dark energy, which *must* be negative gravitational energy in nature. The Big Bang theory is derived from general relativity, which is *entirely* a gravity theory. To require that a gravity theory incorporate a pre-gravity phase within its cosmology, however brief in duration, sounds very much like nonsense. Moreover, cosmic inflation is an ad hoc theory "… contrived with the goal of arranging for the density perturbations to come out right" [Guth (1997), page 238]. Cosmic inflation, in its many different ad hoc forms, appears to be a deeply flawed theory, as nicely elaborated by one of its founders [14].

The purpose of this chapter, however, is not to explain why the FSC model, now integrated into the flat universe Friedmann equations with a cosmological term, rigorously follows observations of cosmic flatness within the CMB. This point has been made in previous FSC publications [Tatum (2015)]. Rather, it is the purpose of this chapter to further explore the possible nature of gravity, dark energy and dark matter. While the FSC model clearly indicates that dark energy is systemic negative gravitational energy, the key question becomes: 'How does this finite constant velocity expanding cosmic system work at its most fundamental level? Specifically, *what is the fundamental nature of its gravity, especially in relation to dark energy and dark matter?*'

Possible clues to the fundamental nature of gravity and dark energy are provided in the new FSC Friedmann equations incorporating a cosmic entropy term. In equation (11) Lambda term Λ is always inversely proportional to the square of the cosmic radius. Thus, Lambda scales downward approximately 121.26 base 10 orders of magnitude from its value during the Planck mass epoch to the current epoch. Interestingly, 10^{121} is the magnitude of the 'cosmological constant problem' [15][16]. Furthermore, equation (11) is seen (in rearranged form) on page 277 of Roger Penrose's latest book [17], if one uses the standard $4\pi R^2$ formula for the cosmic horizon surface area A_{cosm}. Notably, this equation occurs in Penrose's discussion of cosmic entropy, which assumes the Bekenstein-Hawking definition of cosmic entropy [see FSC equation (12)]. So, while Lambda in general relativity is assumed to be a constant by proponents of standard cosmology, the FSC model and Penrose clearly indicate Lambda to be a declining scalar of negative energy density in an expanding *closed* general relativity model. Lambda is clearly an inverse scalar of cosmic entropy, as best seen in FSC equation (13). However, most importantly, as seen by substituting equation (19) into (27B), *total vacuum energy scales in direct proportion to the total entropy term* $\sqrt{S}$. One must keep in mind that the Bekenstein-Hawking definition of cosmic entropy is a unit-less ratio, so $\sqrt{S}$ can also be considered as a measure of cosmic entropy, but on a scale identical to that of the other scaling FSC parameters. **Figures 1 and 2** graphically show the intimate relationship between scaling FSC parameters and total cosmic entropy term $\sqrt{S}$. *It is entirely appropriate to use* $\sqrt{S}$ *as a cosmological clock because equation (18) clearly demonstrates that FSC models the 'entropic arrow of time.'* Notice also that the recently-introduced FSC 'Universal Temperature' T_u [18] inversely scales to the same degree as $\sqrt{S}$ (60.63 logs of 10 from the Planck scale). T_u has a direct one-to-one

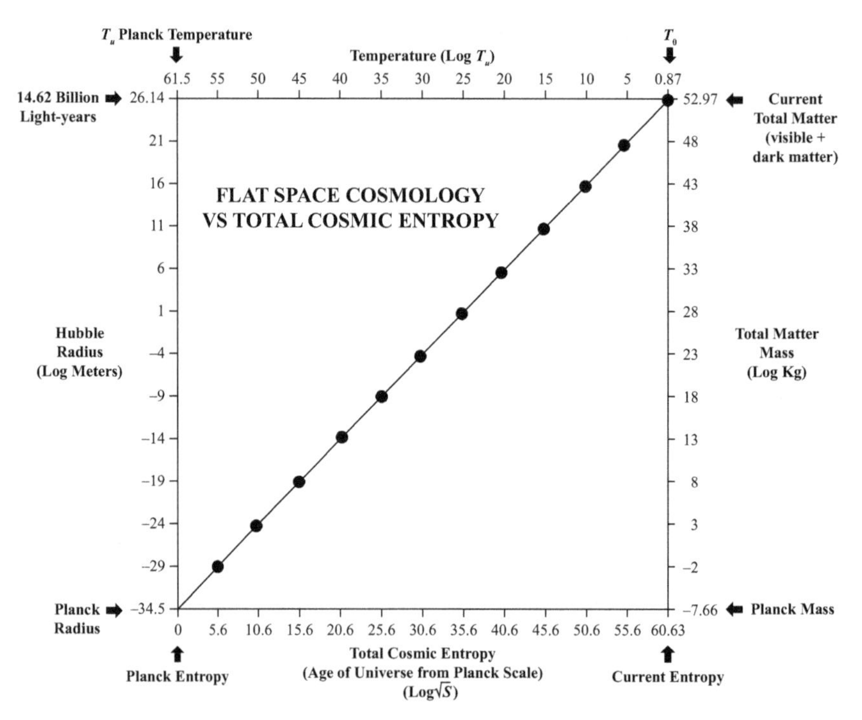

Figure 1 Radius and Total Matter Mass vs Universal Temperature T_u and Entropy.

correspondence to the Kelvin temperature T scale by our Chapter three 'Universal Temperature' definition, $T_u = T^2$.

This idea that total cosmic entropy can be regarded as a cosmological clock is not entirely new, although the FSC model clearly indicates the similarly scaling entropy clock to be in the form of $\sqrt{S}$. Furthermore, the FSC Friedmann entropy equations introduced in this chapter clearly point to cosmic entropy being fundamental to the nature of gravity. Penrose introduces the concept of gravitational entropy to readers on page 256 of his book. Gravitational entropy differs significantly from the entropy of an equilibrated ideal gas, wherein maximum average particle separation at a given temperature characterizes the maximum entropy state. In contrast, *in a gravitating universe, the ongoing clustering of stars and galaxies, and particularly black holes, is in the direction of greater gravitational entropy!* This is made abundantly clear by comparing deep space observational astronomy with observations of (approximately) co-moving galaxies. Supermassive black holes, in particular, are now thought to be huge repositories of total cosmic entropy.

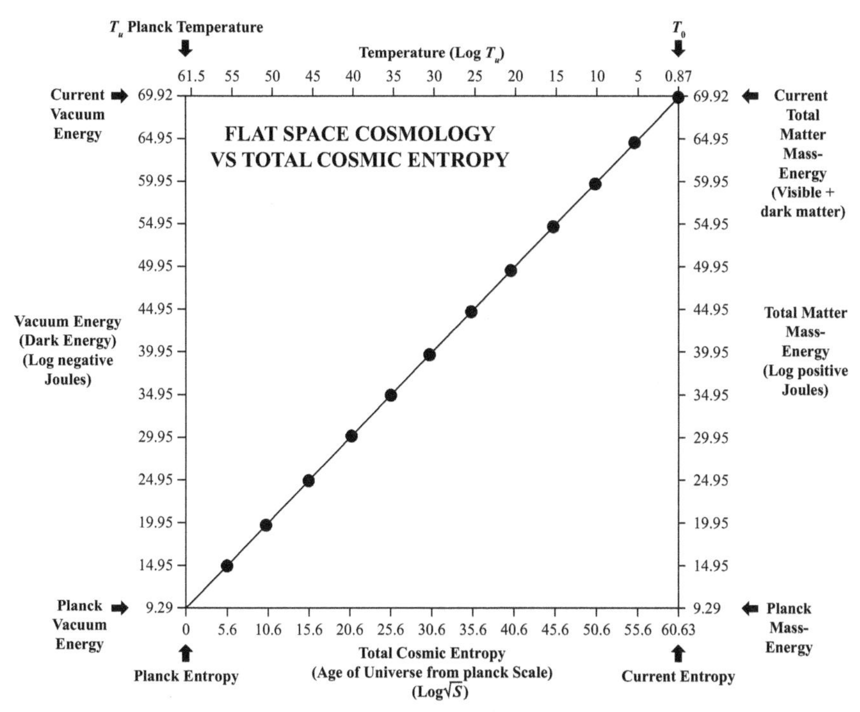

Figure 2 Vacuum Energy and Mass-Energy vs Universal Temperature T_u and Entropy.

A review of the possible fundamental nature of gravity with respect to cosmic entropy should begin with a landmark paper by Erik Verlinde [19]. In this paper, Verlinde makes a persuasive argument that cosmic entropy manifests itself as gravity! He shows in great detail, by a heuristic approach, how gravity could well be an emergent property of cosmic entropy. Perhaps, as suspected by Einstein, gravity and intertia actually follow Mach's Principle, rather than quantum mechanics. Perhaps, at the quantum level, our conventional conception of gravity as a fundamental force might be just as meaningless as a conception of consciousness within two connected neurons. Emergent properties are most evident in complex systems with high degrees of freedom. They can be difficult, if not impossible, to observe or even conceptualize at the smallest scales. This could be a problem for those who assume that gravity should be ultimately definable by rules of quantum physics ('quantum gravity'), unless 'quantum entropy' can somehow be defined. Perhaps the limit of such efforts will be FSC 'quantum cosmology,' as first conceived in 2015 (see Chapter two, Section 2.9).

If Verlinde's compelling emergent property argument ultimately prevails, gravitational inertia (including that of dark matter!) and dark energy could also be emergent properties of cosmic entropy. The total matter mass associated with gravitational entropy, by equations (20) and (24A), must include dark matter. If dark matter observations are somehow a manifestation of unaccounted-for galactic entropy, this might explain why dark matter does not appear to fit within the standard particle model. Alternatively, dark matter might be an overlooked standard model particle, as theorized in Chapters twelve and sixteen.

Verlinde's heuristic approach to a fundamental understanding of gravity as an emergent property of cosmic entropy dovetails nicely with this updated FSC cosmology model. FSC also began as a heuristic model, as did Einstein's (and Planck's) photon. Importantly, neither approach relies in any way on a curved space-time geometrical description of gravity. Although general relativity is a supremely accurate and beautiful theory at large scales, it cannot be a *fundamental* description of gravity. W.S. Krogdahl, for instance, achieved a similarly accurate mathematical model of gravity and cosmology in flat space-time by starting his development with the integration of $E = mc^2$ into Newtonian gravity [20][21][22]. Krogdahl's approach appears to be vindicated by equations (26) thru (29) in this chapter. These equations also give meaning to Newton's discovery that the force of gravity is inversely proportional to R^2!

SUMMARY AND CONCLUSIONS

The purpose of this chapter has been to integrate the highly successful FSC model assumptions, including our entropy assumption, into the Friedmann equations in an effort to explore the fundamental nature of gravity, dark matter and dark energy. In doing so, there should be no doubt that the results are within the rules of general relativity, particularly in light of the fact that black hole equations already follow the rules of general relativity.

The results of this exercise are quite intriguing. The Lambda term Λ in FSC must follow equations (8) and (11), indicating that FSC is a dynamic scalar dark energy (quintessence) model of the wCDM type (the FSC equation of state term w is perpetually -1.0, as shown in Chapter eleven). Lambda is shown to be a declining scalar of negative gravitational vacuum energy density (*i.e.,* dark energy density). Lambda is also an inverse scalar of total cosmic entropy S by equations (13) and (23). Total matter mass-energy, vacuum energy and cosmic time are shown to be directly proportional to total cosmic entropy in the form of $\sqrt{S}$. Thus, the 'entropic arrow of time' is clearly demonstrated in FSC, and $\sqrt{S}$ can be used as the FSC time clock, due to the direct proportionality shown in equation (18) between cosmic entropy $\sqrt{S}$ and cosmic time t (see **Figures 1** and **2**).

A search of recent literature concerning the possible relationship between total cosmic entropy and gravitational interactions identifies Roger Penrose, Stephen Hawking, and Erik Verlinde as pioneers in this field. Penrose's book shows how gravitational entropy clearly differs from the entropy of an ideal gas in the sense that gravitational clustering in the form of stars, galaxies and black holes, is representative of a high (gravitational) entropy state. Black holes, in particular, are now thought to be huge reservoirs of entropy in its highest possible state. In fact, the FSC model, in this context of gravitational entropy, clearly indicates that *black holes may be equivalently defined as localized zones of maximum possible gravitational entropy, from the Planck scale to the scale of the current universe.*

Verlinde's paper on the origin of gravity shows very clearly how gravity could be an emergent property of total cosmic entropy. If so, then gravity may be no more definable at the quantum level than consciousness can be defined within two connected neurons. Moreover, if gravity is truly an emergent property of total cosmic entropy, then it existed from the inception of universal expansion, as opposed to 'freezing out' after a pre-gravity cosmic epoch. This emergent property of entropy concept could also imply that all gravitational manifestations, including gravitational inertia, dark matter and dark energy, are emergent properties of cosmic entropy. If such were the case, these entities might be difficult to define by the rules of quantum physics.

REFERENCES

[1] Tatum, E.T., Seshavatharam, U.V.S. and Lakshminarayana, S. (2015). The Basics of Flat Space Cosmology. International Journal of Astronomy and Astrophysics, 5: 116–124. http://doi.org/10.4236/ijaa.2015.52015

[2] Tatum, E.T., Seshavatharam, U.V.S. and Lakshminarayana, S. (2015). Thermal Radiation Redshift in Flat Space Cosmology. Journal of Applied Physical Science International, 4(1): 18–26.

[3] Tatum, E.T., Seshavatharam, U.V.S. and Lakshminarayana, S. (2015). Flat Space Cosmology as an Alternative to LCDM Cosmology. Frontiers of Astronomy, Astrophysics and Cosmology, 1(2): 98–104. http://pubs.sciepub.com/faac/1/2/3 DOI:10.12691/faac-1-2-3

[4] Tatum, E.T., Seshavatharam, U.V.S. and Lakshminarayana, S. (2015). Flat Space Cosmology as a Mathematical Model of Quantum Gravity or Quantum Cosmology. International Journal of Astronomy and Astrophysics, 5: 133–140. http://doi.org/10.4236/ijaa.2015.53017

[5] Planck Collaboration XIII. (2016). Cosmological Parameters. Astronomy & Astrophysics, 594, A13. DOI:10.1051/0004-6361/201525830. http://arxiv.org/abs/1502.01589

[6] Penrose, R. (1965). Gravitational Collapse and Space-time Singularities. Phys. Rev. Lett., 14: 57.

[7] Hawking, S. and Penrose, R. (1970). The Singularities of Gravitational Collapse and Cosmology. Proc. Roy. Soc. Lond. A, 314: 529–548.

[8] Bekenstein, J.D. (1974). Generalized Second Law of Thermodynamics in Black Hole Physics. Phys. Rev. D, 9: 3292–3300. doi: 10.1103/PhysRevD.9.3292

[9] Hawking, S. (1976). Black Holes and Thermodynamics. Physical Review D, 13(2): 191–197.

[10] Perlmutter, S., et al. (1999). The Supernova Cosmology Project, Measurements of Omega and Lambda from 42 High-Redshift Supernovae. Astrophysical Journal, 517: 565–586. [DOI], [astro-ph/9812133].

[11] Schmidt, B., et al. (1998). The High-Z Supernova Search: Measuring Cosmic Deceleration and Global Curvature of the Universe Using Type Ia Supernovae. Astrophysical Journal, 507: 46–63.

[12] Riess, A.G., et al. (1998). Observational Evidence from Supernovae for an Accelerating Universe and a Cosmological Constant. Astronomical Journal, 116(3): 1009–38.

[13] Guth, Alan. *The Inflationary Universe: The Quest for a New Theory of Cosmic Origins*. New York: Basic Books, 1997.

[14] Steinhardt, P.J. (2011). The Inflation Debate: Is the Theory at the Heart of Modern Cosmology Deeply Flawed? Scientific American, 304(4): 18–25.

[15] Weinberg, S. (1989). The Cosmological Constant Problem. Rev. Mod. Phys., 61: 1–23.

[16] Carroll, S. (2001). The Cosmological Constant. Living Rev. Relativity, 4: 5–56.

[17] Penrose, Roger. (2016). *Fashion, Faith, and Fantasy in the New Physics of the Universe*. Princeton: Princeton University Press.

[18] Tatum, E.T. and Seshavatharam, U.V.S. (2018). Temperature Scaling in Flat Space Cosmology in Comparison to Standard Cosmology. Journal of Modern Physics, 9: 1404–1414. https://doi.org/10.4236/jmp.2018.97085

[19] Verlinde, E. (2010). On the Origin of Gravity and the Laws of Newton. arXiv:1001.0785v1 [hep-th].

[20] Krogdahl, W.S. (2006). Cosmology in Flat Space-Time. arXiv:gr-qc/0402016.

[21] Tatum, E.T. (2017). Why Krogdahl's Flat Space-Time Cosmology is Superior to General Relativity. Journal of Modern Physics, 8: 2087–2095. https://doi.org/10.4236/jmp.2017.813127

[22] Tatum, E.T. (2017). Why the Road to Unification Likely Goes through Krogdahl's Relativity. Journal of Modern Physics, 8: 2096–2103. https://doi.org/10.4236/jmp.2017.813128

CHAPTER 6

How the CMB Anisotropy Pattern Could Be a Map of Gravitational Entropy

Abstract: The rationale for Flat Space Cosmology (FSC) calculations of gravitational entropy in the form of $\sqrt{S}$ is presented. These calculations indicate a tight correlation with the COBE DMR measurement showing CMB RMS temperature variations of 18 micro-Kelvins. The COBE dT/T anisotropy ratio of 0.66×10^{-5} falls within the FSC gravitational entropy range calculated for the beginning and ending conditions of the recombination/decoupling epoch. Thus, an FSC model incorporating gravity as an emergent property of entropy suggests that the CMB temperature anisotropy pattern could simply be a map of gravitational entropy, as opposed to a map of a magnified 'quantum fluctuation' event at a finite beginning of time.*

Keywords: Flat Space Cosmology; Cosmic Microwave Background; CMB Anisotropy; Cosmology Theory; Cosmic Entropy; Gravitational Entropy; Black Holes; Standard Cosmology; Holographic Cosmology

INTRODUCTION AND BACKGROUND

Chapter five makes a persuasive case in support of gravity being an emergent property of cosmic entropy S [1]. This argument is bolstered by Verlinde's landmark paper on the subject [2] and by Roger Penrose's conception of gravitational entropy [3]. Notably, Penrose's presentation on cosmic entropy, which relies on the Bekenstein-Hawking definition of

*Originally published on July 12, 2018 in Journal of Modern Physics (see Appendix refs).

black hole entropy [4][5], relates the magnitude of cosmic horizon surface area ($4\pi R^2$) with the Lambda term Λ in the same way as the Flat Space Cosmology (FSC) model. FSC and Penrose (page 277) have derived Λ as always being equal to $3/R_t^2$. This, of course, implies that vacuum energy density ($\Lambda c^4/8\pi G$) is not a cosmological constant over the great span of cosmic time, but rather a constantly declining cosmological parameter. This relationship applies only to a general relativity model (such as FSC) of an expanding universe with a finite scaling horizon surface area. Only in a *finite* universe model can there a calculable cosmic entropy and a holographic principle ('holographic cosmology').

The key to understanding gravitational entropy, as presented by Penrose (pages 256–258), is that *in a gravitating universe, the ongoing clustering of stars and galaxies is in the direction of greater cosmic entropy*. Black holes, in particular, are thought to be local reservoirs of maximum entropy. If so, then galactic supermassive black holes must be huge repositories of cosmic entropy. In FSC, gravitational entropy in the form of $\sqrt{S}$ scales in direct proportion to FSC cosmic time, cosmic radius, cosmic matter mass, and cosmic vacuum energy (*i.e.,* dark energy). Cosmic entropy in the correct scale form of $\sqrt{S}$ is always inversely proportional to 'Universal Temperature' T_u, as defined by $T_u = T^2$, wherein T^2 is in degrees Kelvin squared. This equal-scaling and proportionality between cosmic gravitational entropy and these FSC parameters allows one to easily calculate the gravitational entropy at any time and temperature in the cosmic past or future. Of particular interest, for the purposes of this chapter, is the *relative* gravitational entropy during the cosmic microwave background (CMB) recombination/decoupling epoch in comparison to the gravitational entropies at one year after the Planck epoch and at current cosmic time in years. Plasma physics and particle physics tell us that the recombination/decoupling event began when our early universe was at about 3,000 K. The great preponderance of the CMB radiation was released during the cosmic time interval extending from when the universe was at 3,000 K to the abrupt 'end of decoupling' approximately 115,000 years later [6].

The astute observer will note that there is some difference between the time vs temperature curves used in standard inflationary cosmology and FSC. These differences are more pronounced in the very early universe. This is the subject of Chapter three [7]. A comparison of these two models in terms of cosmic temperature vs cosmological redshift z is given below. In the FSC model [8][9][10], the following formula is used

$$z \cong \left(\frac{T_t^2}{T_0^2} - 1 \right)^{1/2} \tag{1}$$

wherein T_t is the cosmic radiation temperature at any time t and T_0 is the current observed CMB temperature of 2.72548 K. In standard inflationary cosmology, the following formula is used

$$T_{CMB} \cong 2.725(1+z) \tag{2}$$

wherein T_{CMB} represents the CMB radiation temperature. As presented in Chapter three,

$$T^2 t_{ys} = 1.085781647371578 \times 10^{11} \mathrm{K}^2.\mathrm{yr}\ \text{(sidereal years)} \tag{3}$$

wherein T^2 is in Kelvin squared units and t_{ys} is cosmic time in sidereal years.

In this context, the current chapter analyzes what the FSC model can tell us about the likely effect of gravity on the CMB anisotropy pattern. The implications are discussed in terms of the well-known Sachs-Wolfe effect (See Discussion section below). Particular emphasis is given to the gravitational entropy $\sqrt{S}$ values corresponding to the recombination/decoupling epoch beginning and ending cosmic temperatures. The ensuing discussion will focus on the implications of these gravitational entropy calculations, and what effect gravitational entropy may have had on this CMB anisotropy pattern.

RESULTS

Equation (3) gives an FSC cosmic time value of about 12,064 years at the beginning of the recombination/decoupling epoch (3,000 K). Thus, for reasons given in the Discussion section, the 'end of decoupling' event happened in the FSC model at approximately 127,000 years (924.63 K) after the Planck epoch. The Planck epoch is the time of the Planck-scale universe and is often considered to be the approximate moment of the 'Big Bang' in standard cosmology. Chapter five derives

$$\sqrt{S} = \frac{c\sqrt{\pi}}{L_p} t \tag{4}$$

This shows the direct proportionality relationship between gravitational entropy $\sqrt{S}$ and cosmic time t. Speed of light c and Planck length L_p are assumed to be constants over cosmic time. Thus, as shown in Chapter nine, if we operationally define $\sqrt{S}$ in terms of years,

$\sqrt{S}$ = 1 at 1 year of cosmic time at temperature 3.295×10^5 K
$\sqrt{S}$ = 12,064 at 12,064 years of cosmic time at temperature 3,000 K
$\sqrt{S}$ = 127,000 at 127,000 years of cosmic time at temperature 924.63 K
$\sqrt{S}$ = 14.617×10^9 at 14.617×10^9 years of cosmic time at temperature 2.72548 K

The above CMB gravitational entropies (12,064 and 127,000) can then be related to current cosmic time and entropy (14.617×10^9) as follows:

[$\sqrt{S}$ at the beginning of CMB emission]/[$\sqrt{S}$ at current time] = 8.25×10^{-7} (0.825×10^{-6})
[$\sqrt{S}$ at the ending of CMB emission]/[$\sqrt{S}$ at current time] = 8.69×10^{-6} (0.869×10^{-5})

DISCUSSION

Sachs and Wolfe [11], using a gravitational redshift theoretical argument, suggested that CMB temperature anisotropy could be a result of inhomogeneous gravitational particle clustering *already present* at the beginning of the recombination epoch. Their 1967 gravitational redshift argument for what is known as 'the Sachs-Wolfe effect' is now widely believed to be correct [12]. The Sachs-Wolfe effect is widely considered to be the source of large angular scale temperature fluctuations in the CMB.

However, *in a spatially flat universe, the Sachs-Wolfe effect can also be considered to be the source of the smaller angular scale fluctuations of the CMB temperature anisotropy* [13]. The Boomerang Collaboration [14] reported CMB anisotropy observations closely fitting 'the theoretical predictions for a spatially flat cosmological model with an exactly scale invariant primordial power spectrum for the adiabatic growing mode' [Bucher (2015), page 6]. The Boomerang, Wilkinson Microwave Anisotropy Probe (WMAP) [15], and Planck satellite [16] CMB anisotropy studies have all confirmed global spatial flatness of the universe at the time of the recombination/decoupling epoch. Therefore, in terms of the 'gravitational potential variations' explanation first proposed by Sachs and Wolfe, little in the *theory* of CMB temperature anisotropy has changed since 1967. What *has* changed since the time of the Sachs and Wolfe paper is the *precision of measurements* of the CMB temperature anisotropy. Both the WMAP study and the more sensitive Planck study have confirmed the CMB temperature anisotropy to be on the order of *approximately one part per 100,000* (10^{-5}).

At the time of these CMB study reports, the extreme flatness observations of the CMB temperature anisotropy were credited as a victory for cosmic inflation. However, there was no basis for determining which particular theoretical version of inflation was correct, or even whether another spatial flatness cosmology theory without an inflationary mechanism (such as FSC) could, in fact, be an even better explanation for the CMB observations. The following quote from physicist Philip Gibbs sums it up best: "The problem … is that no particular model of inflation has been shown to work yet. It is possible that work has not yet been completed *or that a more recent specific model will be shown to be right*" [17][18].

As mentioned in the Introduction and Background section, current best estimates of the cosmic time interval during which the CMB radiation was released suggest that the recombination/decoupling epoch lasted approximately 115,000 years. In standard cosmology this is believed to have occurred between approximately 378,000 and 493,000 years after a 'Big Bang' at or near the Planck epoch. In the FSC model, the temperature scaling is noticeably different, placing the beginning of the recombination/decoupling epoch (3,000 K) at approximately 12,064 years after the Planck epoch (see Chapter three). Adding the estimated time interval of approximately 115,000 years puts the FSC 'end of decoupling' event at about 127,000 years after the Planck epoch.

Gravitational entropy $\sqrt{S}$ in the FSC model follows the same log value scale as cosmic time. Thus, there should be a uniform progression from maximum gravitational potential 'smoothness,' corresponding to any operationally-defined 'minimal' or 'beginning' anisotropy, to ongoing and progressively greater gravitational inhomogeneity (*i.e.*, stars, galaxies, galaxy clusters, filaments, black holes and voids). Furthermore, this is consistent with the concept that cosmic entropy smoothly increases as the expanding cosmic horizon surface area (the Bekenstein-Hawking measure of entropy) smoothly increases. Thus, it would seem reasonable to assume that, *if the CMB temperature anisotropy pattern is in keeping with the Sachs-Wolfe effect for a spatially-flat universe, and if gravity is truly an emergent property of cosmic entropy as suggested by Verlinde, the FSC gravitational entropy values pertaining to the recombination/decoupling epoch should also be a measure of the CMB temperature anisotropy.*

The COBE DMR experiment measured CMB RMS temperature variations of 18 micro-Kelvins (1.8×10^{-5} K) [19]. This gives a dT/T anisotropy ratio of (0.000018)/2.725, equaling 6.6×10^{-6} or 0.66×10^{-5}. Little has changed in this respect, judging from the subsequent WMAP and Planck CMB temperature anisotropy findings (also approximately 10^{-5}).

It is intriguing that the FSC gravitational entropy ratios calculated at the end of the Results section are 0.825×10^{-6} at the beginning of the recombination epoch and 0.869×10^{-5} at the 'end of decoupling.' It should be noted that *the 'last scattering surface' is actually a 115,000-year thick segment of microwave radiation spectrum rather than an infinitely thin 'surface' at a single redshift.* In this context, the COBE DMR dT/T anisotropy ratio of 0.66×10^{-5} can only be, in some way, an *averaging* of the actual ratio numbers pertaining to the beginning and ending conditions responsible for the so-called 'last scattering surface.' Therefore, an FSC model incorporating gravity as an emergent property of entropy suggests that the CMB temperature anisotropy pattern could simply be a map of gravitational entropy, as opposed to a map of a magnified 'quantum fluctuation' event at a finite beginning of time.

SUMMARY AND CONCLUSIONS

The purpose of this chapter has been to show how the CMB temperature anisotropy pattern could be a map of gravitational entropy as defined by Roger Penrose in his book entitled *Fashion, Faith and Fantasy in the New Physics of the Universe*. This is particularly relevant with respect to Erik Verlinde's theory that gravity is an emergent property of cosmic entropy. Verlinde's theory dovetails nicely with the July 2018 Journal of Modern Physics paper entitled 'Clues to the Fundamental Nature of Gravity, Dark Energy and Dark Matter' (see Chapter five).

In the present chapter, the rationale for FSC calculations of gravitational entropy in the form of $\sqrt{S}$ is presented. These calculations indicate a tight correlation with the COBE DMR measurement showing CMB RMS temperature variations of 18 micro-Kelvins. The COBE dT/T anisotropy ratio of 0.66×10^{-5} falls within the FSC gravitational entropy range calculated for the beginning and ending conditions of the recombination/decoupling epoch. Thus, an FSC model treating gravity as an emergent property of entropy suggests that the CMB temperature anisotropy pattern could simply be a map of gravitational entropy, as opposed to a map of a magnified 'quantum fluctuation' event at a finite beginning of time.

REFERENCES

[1] Tatum, E.T. and Seshavatharam, U.V.S. (2018). Clues to the Fundamental Nature of Gravity, Dark Energy and Dark Matter. Journal of Modern Physics, 9: 1469–1483. https://doi.org/10.4236/jmp.2018.98091

[2] Verlinde, E. (2010). On the Origin of Gravity and the Laws of Newton. arXiv:1001.0785v1 [hep-th].

[3] Penrose, Roger. *Fashion, Faith, and Fantasy in the New Physics of the Universe*. Princeton: Princeton University Press, 2016.

[4] Bekenstein, J.D. (1974). Generalized Second Law of Thermodynamics in Black Hole Physics. Phys. Rev. D, 9: 3292–3300. DOI: 10.1103/PhysRevD.9.3292

[5] Hawking, S. (1976). Black Holes and Thermodynamics. Physical Review D, 13(2): 191–197.

[6] Spergel, D.N., et al. (2003). First Year Wilkinson Microwave Anisotropy Probe (WMAP) Observations: Determination of Cosmological Parameters. Astrophysical Journal Supplement, 148: 175–194. arXiv:astro-ph/0302209. doi:10.1086/377226

[7] Tatum, E.T. and Seshavatharam, U.V.S. (2018). Temperature Scaling in Flat Space Cosmology in Comparison to Standard Cosmology. Journal of Modern Physics, 9: 1404–1414. https://doi.org/10.4236/jmp.2018.97085

[8] Tatum, E.T., Seshavatharam, U.V.S. and Lakshminarayana, S. (2015). The Basics of Flat Space Cosmology. International Journal of Astronomy and Astrophysics, 5: 116–124. http://doi.org/10.4236/ijaa.2015.52015

[9] Tatum, E.T., Seshavatharam, U.V.S. and Lakshminarayana, S. (2015). Thermal Radiation Redshift in Flat Space Cosmology. Journal of Applied Physical Science International, 4(1): 18–26.

[10] Tatum, E.T., Seshavatharam, U.V.S. and Lakshminarayana, S. (2015). Flat Space Cosmology as an Alternative to LCDM Cosmology. Frontiers of Astronomy, Astrophysics and Cosmology, 1(2): 98–104. http://pubs.sciepub.com/faac/1/2/3. doi:10.12691/faac-1-2-3

[11] Sachs, R.K. and Wolfe, A.M. (1967). Perturbations of a Cosmological Model and Angular Variations of the Microwave Background. Astrophysical Journal, 147: 73. https://doi:10.1086/148982

[12] Wright, E.L. (2003). Theoretical Overview of Cosmic Microwave Background Anisotropy. Carnegie Observatories Astrophysics Series, 2: Measuring and Modeling the Universe. ed. W.L. Freedman. Cambridge University Press, Cambridge, U.K. arXiv:astro-ph/0305591v1.

[13] Bucher, M. (2015). Physics of the Cosmic Microwave Background Anisotropy. arXiv:1501.04288v1 [astro-ph.CO]. https://doi.org/10.1142/S0218271815300049

[14] de Bernardis, P., et al. (2000). A Flat Universe from High-Resolution Maps of the Cosmic Microwave Background Radiation. arXiv:astro-ph/0004404v1. https://doi.org/10.1038/35010035

[15] Bennett, C.L. (2013). Nine-Year Wilkinson Microwave Anisotropy Probe (WMAP) Observations: Final Maps and Results. arXiv:1212.5225v3 [astro-ph.CO]. doi: 10.1088/0067-0049/208/2/20.

[16] Planck Collaboration. (2014). Planck 2013 Results. XXIII. Isotropy and Statistics of the CMB. Astronomy & Astrophysics, A23: 1–48. https://doi.org/10.1015/0004-6361/201321534

[17] Keating, B. (2018). *Losing the Nobel Prize.* W. W. Norton & Company, New York, US.

[18] Gibbs, P.E. (2014). Who Might Get the Nobel Prize for Cosmic Inflation? Prespacetime Journal, 5(3): 230–233. https://prespacetime.com/index.php/pst/article/download/614/612

[19] Wright, E.L., et al. (1996) Angular Power Spectrum of the Cosmic Microwave Background Anisotropy Seen by the COBE DMR. Astrophysical Journal, 464: L21–L24. https://doi.org/10.1086/310073

CHAPTER 7

Predicted Dark Matter Quantitation in FSC

Abstract: The purpose of this chapter is to show how the dark matter predictions of FSC differ with respect to the standard cosmology assertion of a universal dark matter-to-visible matter ratio of approximately 5.3-to-1. FSC predicts the correct ratio to be approximately 9-to-1, based primarily on the universal observations of global spatial flatness in the context of general relativity. The FSC Friedmann equations incorporating a Lambda Λ cosmological term clearly indicate that a spatially-flat universe *must* have equality of the positive curvature (matter mass-energy) and negative curvature (dark energy) density components. Thus, FSC predicts that observations of the Milky Way and the nearly co-moving galaxies within 100 million light years will prove the 5.3-to-1 ratio to be incorrect. The most recent galactic and peri-galactic observations indicate a range of dark matter-to-visible matter ratios varying from essentially zero (NGC 1052-DF2) to approximately 23-to-1. Pending observations promise an exciting next few years for astrophysicists and cosmologists. Within the next few years, the mining of huge data bases (especially the Gaia catalogue and Hubble data) will resolve whether standard cosmology will need to change its current claims for the cosmic energy density partition to be more in line with FSC, or whether FSC is falsified. A prediction is that standard cosmology must eventually realize the *necessity* of resolving the tension between their flatness observations and their assertion of dark energy dominance. It is also a prediction that FSC will soon become the new paradigm in cosmology.*

*Originally published on July 18, 2018 in Journal of Modern Physics (see Appendix refs).

Keywords: Flat Space Cosmology; Standard Cosmology; Cosmology Theory; Dark Matter; Cosmic Microwave Background; Planck Collaboration; Gravitational Entropy; Black Holes

INTRODUCTION AND BACKGROUND

In sharp contrast to standard cosmology, Flat Space Cosmology (FSC) makes quite a number of predictions which would invalidate the theory if proven false. Many of these predictions can be derived from the following FSC Friedmann equation which must always hold true in FSC [1]:

$$\frac{3H^2c^2}{8\pi G} \cong \frac{\Lambda c^4}{8\pi G} \tag{1}$$

wherein the left-hand term is the total matter energy density and the right-hand term is the dark energy density. The H^2 symbol is the squared Hubble parameter value in per seconds squared (s^{-2}) and the Λ symbol is the cosmological parameter in per meters squared (m^{-2}). In a globally spatially-flat expanding universe, *which we observe*, general relativity *stipulates* that the global positive curvature of total matter mass-energy density must exactly offset the global negative curvature of dark energy density. If the case were otherwise, the universe would have a global spatial curvature of sign and magnitude corresponding to whichever happens to be the dominating energy density, *which we do not observe*. Thus, when the universe is at Friedmann's critical density, as appears to be the case by astronomical observations [2], FSC stipulates that 50% of the critical density must be attributable to total matter (visible matter plus dark matter) and 50% of the critical density must be attributable to dark energy.

One of the longstanding observational facts is that the visible matter of our universe comprises only about 5% of the critical density. Thus, FSC predicts a dark matter-to-visible matter ratio of approximately 45-to-5 or 9-to-1. As detailed in the Planck Collaboration consensus report, the ratio of dark matter-to-visible matter is currently claimed to be approximately 5.3-to-1. However, so little is currently known about precisely detecting and quantifying dark matter that *this ratio is subject to higher revision in the likely event that more dark matter is discovered in the future. For this reason, the Planck Collaboration ratio must be considered as a constraint only on the low end.*

Galactic and peri-galactic distributions of dark matter can be surprisingly variable, as evidenced by the 29 March 2018 report in *Nature* [3] of an exceedingly diffuse distant galaxy (NGC 1052-DF2) apparently completely lacking in dark matter! Hence, the *global* (*i.e.,* CMB) Planck Collaboration ratio of 5.3-to-1 cannot dogmatically be considered even an approximation of all galactic and peri-galactic ratios, particularly if these ratios are scalar over cosmic time.

What is now required is a best estimate of the co-moving dark matter-to-visible matter ratio within approximately 50-100 nearby galaxies. The only observable truly co-moving galaxy for us is the Milky Way galaxy itself. All other galaxies are observationally displaced in distance and time to some degree. However, all galaxies within 100 million light years of the Milky Way should be sufficiently close to us to be considered *approximate co-movers* for the required observations. These are the galaxies of the Virgo Supercluster. There are 160 galaxy groups within 100 million light years of the Milky Way galaxy. The number of large galaxies is approximately 2500 and the number of dwarf galaxies is approximately 25,000 [4]. Data mining of the Gaia catalogue will not only allow researchers to determine the *range* of dark matter-to-visible matter ratios within the most accessible nearby co-moving galaxies, but to determine the *average* dark matter-to-visible matter ratio. The average ratio may be calculated by dividing the sum of the dark matter numerators by the total number of nearby co-moving galaxies reliably measured. As mentioned, *FSC predicts this average ratio to be very close to 9-to-1. A radically different average co-moving ratio would falsify FSC, unless there is a significant amount of additional dark matter hidden in deep intergalactic space.* Standard cosmology, on the other hand, has no capacity to predict this ratio. Therefore, whatever this average co-mover ratio turns out to be, it will likely be inserted *ad hoc* into the standard inflationary model *after its determination.*

Given the recent report of the galaxy apparently devoid of dark matter, astrophysicists around the world are scrambling to mine the Gaia catalogue data for further clues with respect to dark matter. The most logical place to start, of course, is with the Milky Way galaxy. Remarkably, this data has just become available! This author predicts that Posti and Helmi's May 2018 arXiv.org publication of 'Mass and Shape of the Milky Way's Dark Matter Halo with Globular Clusters from Gaia and Hubble' [5] will be considered a landmark publication concerning galactic and peri-galactic dark matter. *This study reveals that the virial volume comprising our Milky Way Galaxy and its dark matter halo has a dark matter-to-visible matter ratio of approximately 23.074-to-1. Thus, the matter confined within the halo radius ('virial radius') of our Milky Way galaxy appears to be approximately 95.85% dark matter and 4.15% visible matter!*

So that the reader can make the same calculations, the relevant measurements made by Posti and Helmi are repeated here: the virial mass is reported to be 1.3 +/– 0.3 $\times 10^{12}$ solar masses; the mass of the Milky Way galaxy within a generous 20 kpc radius (a radius of approximately 65,200 light-years) is reported to be 1.91×10^{11} solar masses, of which Posti and Helmi attribute 1.37×10^{11} solar masses to dark matter. One can, therefore, assume the remaining 0.54×10^{11} solar masses to be the galactic visible matter within approximately 65,200 light-years of the Milky Way center. This can safely be assumed to be greater than 95% of the Milky Way visible matter. This is because numerous reliable sources indicate the visible matter of the Milky Way to be within a radius of about 50,000 light-years of the galactic center. Posti and Helmi measured a virial radius of 287 kpc. This is greater than 14 times their defined radius of the galactic disc, and surely must encapsulate the vast majority of the Milky Way dark matter. Hence, one can assume that the ratio of 1.3×10^{12} solar masses to 0.54×10^{11} solar masses is an excellent approximation of the Milky Way dark matter-to-visible matter ratio. This is how the author calculated the ratio and percentage numbers in the prior paragraph.

These very recent observations ranging from 0% galactic dark matter (NGC 1052-DF2) to greater than 95% dark matter within the virial volume of the Milky Way galaxy must be somewhat jarring to standard model proponents. The current assertion of approximately 30% universal total matter mass-energy and approximately 70% dark energy appears to be on a shaky foundation [6][7][8][9][10][11]. Within the next few years, the mining of huge data sets (especially the Gaia catalogue and Hubble data) will resolve whether standard cosmology will need to change its current claims for the cosmic energy density partition to be more in line with FSC, or whether FSC is falsified. Regardless, standard cosmology must eventually realize the *necessity* of resolving the tension between their flatness observations and their current assertion of dark energy dominance.

SUMMARY AND CONCLUSIONS

The purpose of this chapter has been to show how the dark matter predictions of FSC differ with respect to the standard cosmology assertion of a universal dark matter-to-visible matter ratio of approximately 5.3-to-1. FSC predicts the correct ratio to be approximately 9-to-1, based primarily on the universal observations of global spatial flatness in the context of general relativity. The FSC Friedmann equations incorporating a Lambda Λ cosmological term clearly indicate that a spatially-flat universe *must* have equality of the global positive curvature (matter mass-energy) and global negative curvature (dark energy) density components. Thus, FSC predicts that observations of the Milky Way and the nearly co-moving galaxies within 100 million light years will prove the 5.3-to-1 ratio to be incorrect.

The most recent galactic observations indicate a range of dark matter-to-visible matter ratios varying from essentially zero (NGC 1052-DF2) to approximately 23-to-1. Pending observations promise an exciting next few years for astrophysicists and cosmologists. Within the next few years, the mining of huge data bases will resolve whether standard cosmology will need to change its current claims for the cosmic energy density partition to be more in line with FSC, or whether FSC is falsified. A prediction is that standard cosmology must eventually realize the *necessity* of resolving the tension between their flatness observations and their assertion of dark energy dominance. It is also a prediction that FSC will soon become the new paradigm in cosmology.

REFERENCES

[1] Tatum, E.T. and Seshavatharam, U.V.S. (2018). Clues to the Fundamental Nature of Gravity, Dark Energy and Dark Matter. Journal of Modern Physics, 9: 1469–1483. https://doi.org/10.4236/jmp.2018.98091

[2] Planck Collaboration XIII. (2016). Cosmological Parameters. Astronomy & Astrophysics, 594, A13. doi:10.1051/0004-6361/201525830 http://arxiv.org/abs/1502.01589

[3] van Dokkum, P., et al. (2018). A Galaxy Lacking Dark Matter. Nature, 555: 629–632. doi:10.1038/nature25767

[4] Theuns, T. (2003). The Universe within 100 Million Light Years. Institute of Computational Cosmology. www.icc.dur.ac.uk/~tt/Lectures/Galaxies/LocalGroup/Back/virgo.html

[5] Posti, L. and Helmi, A. (2018). Mass and Shape of the Milky Way's Dark Matter Halo with Globular Clusters from Gaia and Hubble. arXiv:1805.01408v1 [astro-ph.GA].

[6] Tatum, E.T. and Seshavatharam, U.V.S. (2018). How a Realistic Linear R_h = ct Model of Cosmology Could Present the Illusion of Late Cosmic Acceleration. Journal of Modern Physics, 9: 1397–1403. https://doi.org/10.4236/jmp.2018.97084

[7] Tutusaus, I., et al. (2017). Is Cosmic Acceleration Proven by Local Cosmological Probes? Astronomy & Astrophysics, 602_A73. arXiv:1706.05036v1 [astro-ph.CO]. doi:10.1051/0004-6361/201630289

[8] Dam, L.H., et al. (2017). Apparent Cosmic Acceleration from Type Ia Supernovae. Mon. Not. Roy. Astron. Soc. arXiv:1706.07236v2 [astro-ph.CO]. doi:10.1093/mnras/stx1858

[9] Nielsen, J.T., et al. (2015). Marginal Evidence for Cosmic Acceleration from Type Ia Supernovae. Scientific Reports, 6: Article number 35596. arXiv:1506.01354 [astro-ph.CO]. doi:10.1038/srep35596

[10] Jun-Jie Wei, et al. (2015). A Comparative Analysis of the Supernova Legacy Survey Sample with ΛCDM and the R_h = ct Universe. *Astronomical Journal,* 149: 102–113. doi:10.1088/0004-6256/149/3/102

[11] Melia, F. (2012). Fitting the Union 2.1 SN Sample with the Rh = ct Universe. Astronomical Journal, 144: arXiv:1206.6289 [astro-ph.CO] doi:10.1088/0004-6256/144/4/110

CHAPTER 8

Flat Space Cosmology as a Model of Penrose's Weyl Curvature Hypothesis and Gravitational Entropy

Abstract: FSC is shown to be an excellent model of Penrose's Weyl curvature hypothesis and his concept of gravitational entropy. The assumptions of FSC allow for the minimum entropy at the inception of the cosmic expansion and rigorously define a cosmological arrow of time. This is in sharp contrast to inflationary models, which appear to violate the second law of thermodynamics within the early universe. Furthermore, by virtue of the same physical assumptions applying at any cosmic time t, the perpetually-flat FSC model predicts the degree of scale invariance observed in the CMB anisotropy pattern, without requiring an explosive and exceedingly brief inflationary epoch. Penrose's concepts, as described in this chapter, provide support for the idea that FSC models gravitational entropy and Verlinde's emergent gravity theory.*

Keywords: Flat Space Cosmology; Cosmology Theory; Gravitational Entropy; Weyl's Curvature Hypothesis; Black Holes; Cosmic Inflation; Cosmic Flatness; Cosmic Microwave Background

---

*Originally published on September 4, 2018 in Journal of Modern Physics (see Appendix refs).

INTRODUCTION AND BACKGROUND

If the expanding universe follows the second law of thermodynamics, then the total cosmic entropy of each earlier epoch in time *must* have had a lower value. The various theories of cosmic inflation appear to ignore this stipulation, as detailed by Roger Penrose in his 2016 book entitled *Fashion, Faith and Fantasy in the New Physics of the Universe* [1]. Penrose makes a convincing argument that the *gravitational* entropy state of the earliest universe (*i.e., before* the supposed inflationary epoch at 10^{-37} to 10^{-32} second of cosmic time) must have been exceedingly low, even in comparison to the post-inflationary universe. This is in stark contrast to the belief on the part of inflationists that inflation 'solves' the problem posed by an extremely chaotic (*i.e.,* high entropy) Big Bang quantum fluctuation event as a beginning of the universe. Yet, as Penrose points out, any such inflationary solution would be either a violation of the second law or an unnecessary solution to a non-existent problem (were inflation to begin in a smooth flat patch of primordial space-time). So, in Penrose's view, a theory of cosmic inflation is either a violation of physics or, at best, completely unnecessary. Others have expressed similar concerns with inflationary theory, including one of its founders [2].

In 1979, before any theories of inflation were proposed, Penrose first addressed the tension between the remarkable apparent homogeneity and isotropy of the universe (also inherent in the FLRW model) and the second law requirement of extremely low beginning entropy. He introduced the concept of 'gravitational entropy,' wherein the ongoing clustering of stars and galaxies, and the formation of black holes, is in the direction of ever-greater total cosmic entropy. Thus, in stark contrast to the thermal entropy of a gas, the total entropy of a gravitating system can be considered to be lowest at its smallest scale and its most homogeneous gravitational state. In his 'Weyl curvature hypothesis' [3], Penrose associates the lowest entropy state of the earliest universe with a vanishing Weyl curvature tensor. Thus, if one wishes to consider the theoretical possibility of a 'Big Bang' from a singularity condition, under this hypothesis, the universe begins 'free of independent gravitational degrees of freedom' [Penrose (2016), pages 371–374]. Operationally, one can consider this to be a 'zero' gravitational entropy state with respect to any future positive gravitational entropy values of the expanding universe. This gravitational entropy concept is a key feature of the Flat Space Cosmology (FSC) model [4] and Erik Verlinde's 'emergent gravity' (entropic gravity) theory [5][6]. The concept of gravitational entropy also allows for a very tight FSC correlation with the observed anisotropy magnitude (0.66×10^{-5} dt/T) in the cosmic microwave background (CMB) [7].

The Weyl curvature hypothesis and the FSC model both provide for a cosmological arrow of time [8]. This is in contrast to the standard inflationary model, which has no effective means by which the law of entropy can be obeyed if the model starts with a highly chaotic and high entropy quantum fluctuation event.

One of the current sources of tension between the most recent CMB observations [9] and standard inflationary cosmology is the assertion of global cosmological spatial flatness and dark energy dominance. The assertion of flatness stipulates a Friedmann curvature k term value of zero, while the assertion of dark energy dominance stipulates a small negative value to the k term. *Both stipulations cannot be true at the same time!* Furthermore, while the near scale invariance of the CMB power spectrum is commonly touted to be a validation of inflationary cosmology, scale invariance is even easier to explain in FSC. Scale invariance in a cosmological model essentially means that the same laws of physics apply to any scale of the cosmological model. Thus, one would *expect* a large degree of self-similarity between adjacent cosmological epochs in a spatially-flat model in which the same set of basic assumptions are prescribed to occur at any cosmological time t. The following section presents the current five basic assumptions of FSC.

THE FIVE BASIC ASSUMPTIONS OF FSC

FSC models the Hawking-Penrose conjecture that a smoothly-expanding cosmic system beginning from a singularity can be modeled *within the rules of general relativity* as a time-reversed black hole. Hence, the assumptions of FSC are as follows:

1. The cosmic model is an ever-expanding sphere such that the cosmic horizon always translates at speed of light c with respect to its geometric center at all times t. The observer is operationally-defined to be at this geometric center at all times t.

2. The cosmic radius R_t and total mass M_t follow the Schwarzschild formula $R_t \cong 2GM_t/c^2$ at all times t.
3. The cosmic Hubble parameter is defined by $H_t \cong c/R_t$ at all times t.
4. Incorporating our cosmological scaling adaptation of Hawking's black hole temperature formula, at any radius R_t, cosmic temperature T_t is inversely proportional to the geometric mean of cosmic total mass M_t and the Planck mass M_{pl}. R_{pl} is defined as twice the Planck length (*i.e.*, as the Schwarzschild radius of the Planck mass black hole). With subscript t for any time stage of cosmic evolution and subscript pl for the Planck scale epoch, and, incorporating the Schwarzschild relationship between M_t and R_t,

$$\left.\begin{array}{ll} k_B T_t \cong \dfrac{\hbar c^3}{8\pi G\sqrt{M_t M_{pl}}} \cong \dfrac{\hbar c}{4\pi\sqrt{R_t R_{pl}}} & \\ \left\{\begin{array}{ll} M_t \cong \left(\dfrac{\hbar c^3}{8\pi G k_B T_t}\right)^2 \dfrac{1}{M_{pl}} & \text{(A)} \\ R_t \cong \dfrac{1}{R_{pl}}\left(\dfrac{\hbar c}{4\pi k_B}\right)^2 \left(\dfrac{1}{T_t}\right)^2 & \text{(B)} \\ R_t T_t^2 \cong \dfrac{1}{R_{pl}}\left(\dfrac{\hbar c}{4\pi k_B}\right)^2 & \text{(C)} \\ t \cong \dfrac{R_t}{c} & \text{(D)} \end{array}\right. & \end{array}\right\} \quad (1)$$

5. Total entropy S_t of the cosmic model follows the Bekenstein-Hawking black hole formula [10][11], wherein R_t is the cosmic radius at time t and L_p is the Planck length.

$$S_t \cong \frac{\pi R_t^2}{L_p^2} \quad (2)$$

These model assumptions correlate very closely with current observations, as detailed in the first two chapters It should be remembered that all five of these assumptions apply to every second of the FSC cosmological model. Hence, *scale invariance to the degree seen in the CMB anisotropy pattern is a prediction of FSC and must not be considered the exclusive domain of inflationary models.*

PERPETUAL FRIEDMANN'S CRITICAL DENSITY IN FSC

As described in some detail in the seminal FSC papers [12][13][14][15], the first three assumptions allow for perpetual Friedmann's critical density (*i.e.*, perpetual global spatial flatness) of the expanding FSC cosmological model from inception. By dividing the Schwarzschild mass (defined in terms of cosmic radius R_0) by the spherical volume, and substituting c^2/R_0^2 with H_0^2, Friedmann's critical mass density $\rho_0 = 3H_0^2/8\pi G$ is achieved for any given moment of observation (hence the subscript '0') in cosmic time. So, *perpetual Friedmann's critical density (i.e., perpetual spatial flatness) from inception is a fundamental feature of the FSC model.* And the current observational *global* Hubble parameter H_0 value is calculated by the FSC model to be 66.9 km.s^{-1}.Mpc^{-1}, which fits the lower end range of the of the 2015 Planck Collaboration consensus observational value of 67.8 +/– 0.9 km.s^{-1}.Mpc^{-1} (68% confidence interval), as shown in Chapter one.

GRAVITATIONAL ENTROPY IN FSC

FSC models the Hawking-Penrose conjecture that a smoothly-expanding cosmic system beginning from a singularity can be modeled *within the rules of general relativity* as a time-reversed black hole. Thus, assumption #5 defining FSC entropy by $S_t \cong \pi R_t^2 / L_p^2$ at all times t seems appropriate. There appear to be no known scale limitations of a cosmological model which follows the above assumptions. And yet, each successively earlier stage of the FSC model *must* have a lower (gravitational) entropy value by virtue of the assumption that the entropy of a black hole scales according to R_t^2. Thus, it is shown by the above theoretical considerations and model assumptions that FSC is a model of Penrose's Weyl curvature hypothesis, his concept of gravitational entropy, and Verlinde's theory of emergent gravity.

SUMMARY AND CONCLUSIONS

FSC has been shown to be an excellent model of Penrose's Weyl curvature hypothesis and his concept of gravitational entropy. The assumptions of FSC allow for the minimum entropy at the inception of the cosmic expansion and rigorously define a cosmological arrow of time. This is in sharp contrast to inflationary models, which appear to violate the second law of thermodynamics within the early universe. Furthermore, by virtue of the same physical assumptions applying at any cosmic time t, the perpetually-flat FSC model predicts the degree of scale invariance observed in the CMB anisotropy pattern, without requiring an explosive and exceedingly brief inflationary epoch. Penrose's concepts, as described in this chapter, provide support for the idea that FSC models gravitational entropy and Verlinde's emergent gravity theory.

REFERENCES

[1] Penrose, Roger. (2016). *Fashion, Faith, and Fantasy in the New Physics of the Universe*. Princeton: Princeton University Press.

[2] Steinhardt, P.J. (2011). The Inflation Debate: Is the Theory at the Heart of Modern Cosmology Deeply Flawed? Scientific American, 304(4): 18–25.

[3] Penrose, R. (1979). Singularities and Time-Asymmetry. In S.W. Hawking and W. Israel. *General Relativity: An Einstein Centenary Survey*. Cambridge University Press. pp. 581–638.

[4] Tatum, E.T. and Seshavatharam, U.V.S. (2018). Clues to the Fundamental Nature of Gravity, Dark Energy and Dark Matter. Journal of Modern Physics, 9: 1469–1483. https://doi.org/10.4236/jmp.2018.98091

[5] Verlinde, E. (2010). On the Origin of Gravity and the Laws of Newton. arXiv:1001.0785v1 [hep-th].

[6] Verlinde, E. (2016). Emergent Gravity and the Dark Universe. aeXiv: 1611.02269v2 [hep-th].

[7] Tatum, E.T. (2018). How the CMB Anisotropy Pattern Could Be a Map of Gravitational Entropy. Journal of Modern Physics, 9: 1484–1490. https://doi.org/10.4236/jmp.2018.98092

[8] Tatum, E.T. and Seshavatharam, U.V.S. (2018). Temperature Scaling in Flat Space Cosmology in Comparison to Standard Cosmology. Journal of Modern Physics, 9: 1404–1414. https://doi.org/10.4236/jmp.2018.97085

[9] Planck Collaboration XIII. (2016). Cosmological Parameters. Astronomy & Astrophysics, 594, A13. doi: 10.1051/0004-6361/201525830. http://arxiv.org/abs/1502.01589

[10] Bekenstein, J.D. (1974). Generalized Second Law of Thermodynamics in Black Hole Physics. Phys. Rev. D, 9: 3292–3300. doi: 10.1103/PhysRevD.9.3292

[11] Hawking, S. (1976). Black Holes and Thermodynamics. Physical Review D, 13(2): 191–197.

[12] Tatum, E.T., Seshavatharam, U.V.S. and Lakshminarayana, S. (2015). The Basics of Flat Space Cosmology. International Journal of Astronomy and Astrophysics, 5: 116–124. http://doi.org/10.4236/ijaa.2015.52015

[13] Tatum, E.T., Seshavatharam, U.V.S. and Lakshminarayana, S. (2015). Thermal Radiation Redshift in Flat Space Cosmology. Journal of Applied Physical Science International, 4(1): 18–26.

[14] Tatum, E.T., Seshavatharam, U.V.S. and Lakshminarayana, S. (2015). Flat Space Cosmology as an Alternative to LCDM Cosmology. Frontiers of Astronomy, Astrophysics and Cosmology, 1(2): 98–104. http://pubs.sciepub.com/faac/1/2/3 doi: 10.12691/faac-1-2-3

[15] Tatum, E.T., Seshavatharam, U.V.S. and Lakshminarayana, S. (2015). Flat Space Cosmology as a Mathematical Model of Quantum Gravity or Quantum Cosmology. International Journal of Astronomy and Astrophysics, 5: 133–140. http://doi.org/10.4236/ijaa.2015.53017

CHAPTER 9

Calculating Radiation Temperature Anisotropy in Flat Space Cosmology

Abstract: The purpose of this chapter is to show how one can use the FSC model of gravitational entropy to calculate cosmic radiation temperature anisotropy for any past cosmic time t since the Planck scale epoch. Cosmic entropy follows the Bekenstein-Hawking definition, although in the correct-scaling form of $\sqrt{S}$, which scales 60.63 logs of 10 from the Planck scale. In the FSC model, cosmic radiation temperature anisotropy $A_t = (t/t_0)$. The derived past anisotropy value can be compared to current co-moving anisotropy defined as unity (t_0/t_0). Calculated in this way, current gravitational entropy and temperature anisotropy have maximum values, and the earliest universe has the lowest entropy and temperature anisotropy values. This approach comports with the second law of thermodynamics and the theoretical basis of the Sachs-Wolfe effect, gravitational entropy as defined by Roger Penrose, and Erik Verlinde's 'emergent gravity' theory.*

Keywords: Flat Space Cosmology; Cosmic Microwave Background; CMB Anisotropy; Cosmology Theory; Cosmic Entropy; Gravitational Entropy; Black Holes; Standard Cosmology

INTRODUCTION AND BACKGROUND

Chapter six [1] presents the rationale for Flat Space Cosmology (FSC) calculations of gravitational entropy in the form of $\sqrt{S}$. The theoretical

---

*Originally published on September 4, 2018 in Journal of Modern Physics (see Appendix refs).

basis for doing so is the Sachs-Wolfe effect [2]. The Sachs-Wolfe effect is widely considered to be the source of large angular scale temperature fluctuations in the cosmic microwave background (CMB). However, in a spatially flat universe, the Sachs-Wolfe effect can also be considered to be the source of the smaller angular scale fluctuations of the CMB temperature anisotropy [3]. The Boomerang Collaboration [4] reported CMB anisotropy observations closely fitting 'the theoretical predictions for a spatially flat cosmological model with an exactly scale invariant primordial power spectrum for the adiabatic growing mode' [Bucher (2015), page 6]. Thus, the Sachs-Wolfe effect as a measure of 'gravitational potential' appears to be the basis for tight correlation between the FSC CMB anisotropy calculations and the observed CMB anisotropy [5][6][7].

Furthermore, the FSC CMB anisotropy paper [1] (Chapter six) and its companion paper [8] (Chapter five) show how the FSC model dovetails nicely with Erik Verlinde's concept of 'emergent gravity' as an emergent property of cosmic entropy [9][10] and Roger Penrose's concept of gravitational entropy [11] based upon his 'Weyl's curvature hypothesis' [12][13].

The purpose of this chapter is to show how the FSC CMB anisotropy concept opens the way for a definition of cosmic radiation temperature anisotropy *at any cosmic temperature T*, whether it is in the form of 'Universal Temperature' T_U, as defined in Chapter three [14], or in the Kelvin temperature scale. Although Chapter six shows how to calculate the gravitational entropy ratios relating entropies at given *years* of cosmic time since the Planck epoch, these values can also be calculated in given *seconds* of cosmic time since the Planck epoch. This allows for the first-year cosmic times and gravitational entropies to be correlated second-by-second as shown below [see equation (1)]. Therefore, cosmic radiation temperature anisotropy is not only predicted by FSC for the CMB recombination/decoupling 'last scattering surface' but also for *any other cosmic time t*.

The rationale for generalizing the radiation temperature anisotropy calculations to be presented herein can be summarized as follows: Sachs and Wolfe used a gravitational redshift theoretical argument that radiation temperature anisotropy could be a result of inhomogeneous gravitational particle clustering. So, while they applied their argument in anticipation of refined CMB anisotropy observations, *there was nothing particularly special about the CMB emission event with respect to their gravitational redshift argument.* The recombination/decoupling event concerned photon emission, making the event observable. However, the coupling of electrons with protons to form the first hydrogen atoms should have no impact whatsoever on gravitational particle clustering. As explained in Chapter six, the Sachs-Wolfe effect is now widely considered to be the theoretical basis for large angular scale radiation temperature fluctuations. Furthermore, *in a spatially flat universe, the Sachs-Wolfe effect can also be considered to be the source of the smaller angular scale fluctuations of the CMB temperature anisotropy [3].* Therefore, based upon this scale-invariant flat universe Sachs-Wolfe effect, the calculation method presented below is believed to be generalizable to any other gravitational particle clustering stage (*i.e.*, gravitational entropy stage) of universal expansion.

The reader will maximally benefit in reading the present chapter after first reviewing Chapters one, three, five and six.

CALCULATING RADIATION TEMPERATURE ANISOTROPY IN FSC

Equation (4) from Chapter three gives an FSC cosmic time value of $10^{11.58}$ seconds at the beginning of the recombination/decoupling epoch (3,000 K). This equation is repeated here as equation (1):

$$T^2 t_s = 3.426525959553982 \times 10^{18}\,\mathrm{K}^2.\mathrm{s} \tag{1}$$

wherein cosmic temperature T is in degrees K and cosmic time t_s is in seconds since the Planck epoch. For reasons given in the Chapter six discussion section, the 'end of decoupling' event happened in the FSC model at $10^{12.6}$ seconds (at a temperature of 924.63 K) after the Planck epoch. The Planck epoch is the cosmic time of the Planck-scale universe and is often considered to be the approximate moment of the 'Big Bang' in standard cosmology. Chapter five derives

$$\sqrt{S} - \frac{c\sqrt{\pi}}{L_p} t \tag{2}$$

Showing the direct proportionality relationship between gravitational entropy $\sqrt{S}$ and cosmic time t. Speed of light c and Planck length L_p are assumed to be constants over cosmic time. Thus, if we normalize the proportionality constant to unity and operationally define $\sqrt{S}$ in terms of seconds,

$\sqrt{S} = 10^{-42.965}$ at $10^{-42.965}$ second of cosmic time at Planck temperature 5.6×10^{30} K

$\sqrt{S} = 10^{11.58}$ at $10^{11.58}$ seconds of cosmic time at CMB beginning temperature 3,000 K

$\sqrt{S} = 10^{12.6}$ at $10^{12.6}$ seconds of cosmic time at end-of-decoupling temperature 924.63 K

$\sqrt{S} = 10^{17.66}$ at $10^{17.66}$ seconds of cosmic time at current temperature 2.72548 K

The above CMB emission epoch gravitational entropies ($10^{11.58}$ and $10^{12.6}$) can then be related to current cosmic entropy ($10^{17.66}$), in ratio form, as follows:

[$\sqrt{S}$ at the beginning of CMB emission]/[$\sqrt{S}$ at current time] $= 8.25 \times 10^{-7}$ (0.825×10^{-6})

[$\sqrt{S}$ at the ending of CMB emission]/[$\sqrt{S}$ at current time] $= 8.69 \times 10^{-6}$ (0.869×10^{-5})

These derived values of the beginning and ending CMB radiation temperature anisotropy are the same as those calculated in Chapter six [1]. This value range fits the COBE DMR CMB dT/T anisotropy measurement of 0.66×10^{-5}, as well as the WMAP and Planck report anisotropy estimates of 'approximately' 10^{-5}.

Thus, if this process for calculating the CMB radiation temperature anisotropy as gravitational entropy ratios can be generalized and extended all the way back to the Planck epoch (and conceivably beyond), the formula for doing such calculations is:

$$A_t = \sqrt{S_t} / \sqrt{S_o} = t / t_o \qquad (3)$$

wherein A_t is the radiation temperature anisotropy at cosmic time t, $\sqrt{S_t}$ is the gravitational entropy (cosmic entropy) at cosmic time t, $\sqrt{S_0}$ is the gravitational entropy (cosmic entropy) at current time t_0, and t/t_0 is the ratio of these time values. This radiation temperature anisotropy formula can also be substituted by other FSC parameters correlated to $\sqrt{S_t}$, as seen in Chapter five. Thus,

$$A_t = R_t / R_0 \quad (4)$$

wherein R_t is the cosmic radius at cosmic time t and R_0 is the current observed cosmic radius.

$$A_t = T_0^2 / T_t^2 \quad (5)$$

wherein T_0 is the current cosmic radiation temperature in degrees Kelvin (2.72548 K) and T_t is the cosmic radiation temperature in degrees Kelvin at cosmic time t.

$$A_t = T_{U0} / T_{Ut} \quad (6)$$

wherein T_{U0} is the current cosmic radiation temperature in Universal Temperature units and T_{Ut} is the cosmic radiation temperature in Universal Temperature units at cosmic time t. As indicated in Chapter three [14], T_{Ut} is defined by $T_{Ut} = T_t^2$.

Based upon this calculation method, **Figures 1** and **2** can be presented. For comparison, the reader is referred to Figures 1 and 5 in Chapter three.

Figure 1 shows how radiation temperature anisotropy and total cosmic entropy in the form of $\sqrt{S}$ scales with respect to cosmic time t and Kelvin temperature T. CMB starting and ending values are denoted by the white circles.

Figure 2 shows how radiation temperature anisotropy and total cosmic entropy in the form of $\sqrt{S}$ scales with respect to cosmic time t and Universal Temperature T_U. CMB starting and ending values are denoted by the white circles.

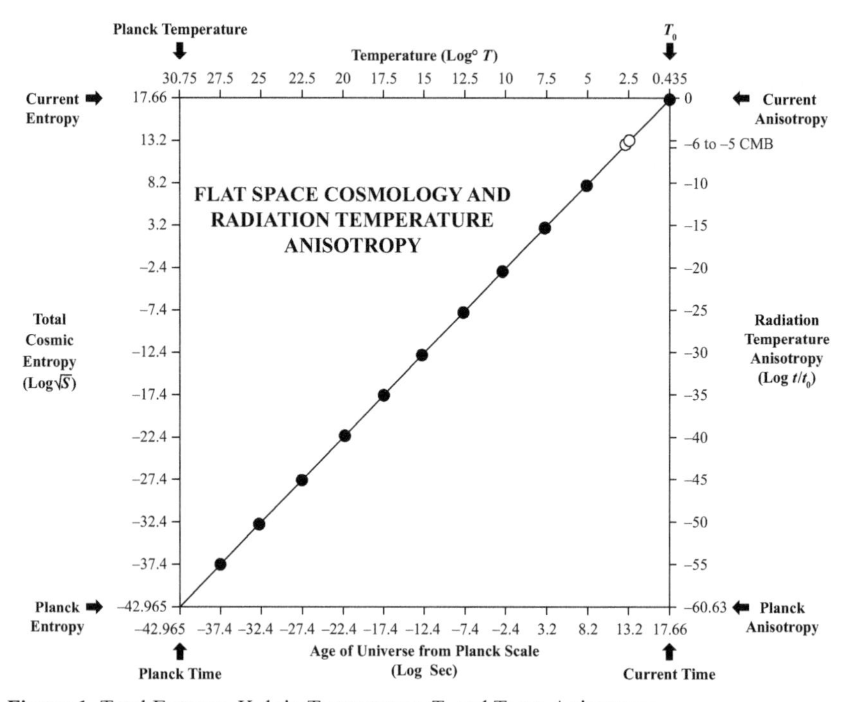

Figure 1 Total Entropy, Kelvin Temperature T, and Temp Anisotropy.

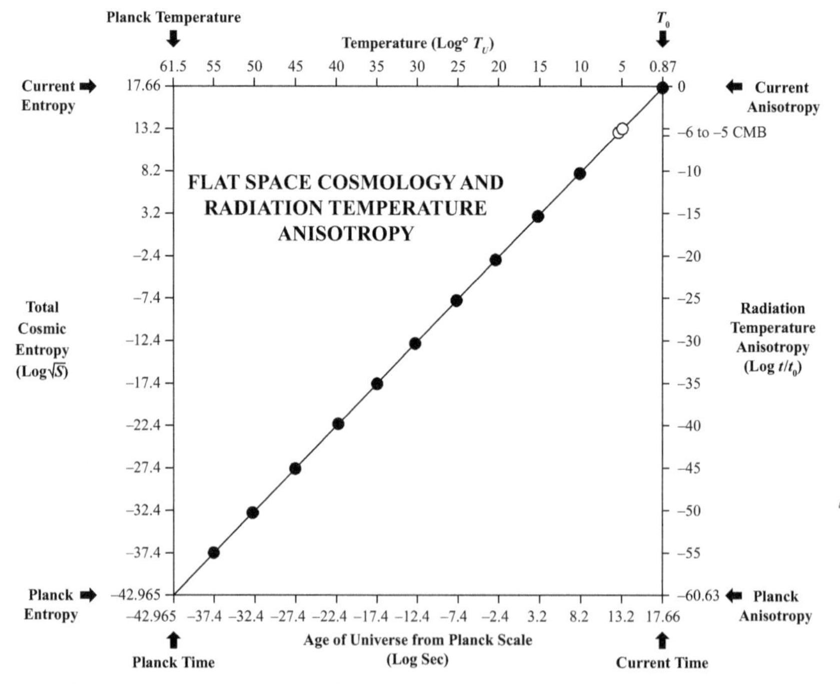

Figure 2 Total Entropy, Universal Temperature T_U, and Temp Anisotropy.

DISCUSSION

Extensive comparisons between FSC and standard inflationary cosmology are given in Chapter two [15]. Standard inflationary cosmology has no theoretical basis for predicting and calculating radiation temperature anisotropy at any given cosmic time. Thus, standard model practitioners can only guess as to the possible reasons why the CMB anisotropy pattern has a *dT/T* RMS anisotropy value of 18 micro-Kelvins/2.725 K = 0.66×10^{-5}, as measured in the COBE DMR experiment [5]. Current speculation seems to favor a CMB pattern produced by a 'quantum fluctuation' Big Bang event smoothed out by cosmic inflation and splayed out across the sky. It is even proposed that a 'quantum fluctuation' CMB pattern somehow must provide important clues to the nature of gravity at the quantum scale (*i.e.*, 'quantum gravity').

In contrast, the FSC model, as detailed in Chapters five [8] and six [1], indicates that the CMB pattern could simply be a map of gravitational entropy. The theoretical basis for this interpretation owes much to the prior work of Sachs and Wolfe, Hawking, Penrose and Verlinde, as discussed in both chapters. Verlinde's papers [9] [10], in particular, address the deep correlation between cosmic entropy and gravity. Thus, given the FSC success in correlating gravitational entropy in the form of $\sqrt{S}$ with the observed anisotropy of the CMB pattern, the current paper proposes a reasonable extension of the same rationale to calculating radiation temperature anisotropy at any past cosmic time t relative to current cosmic anisotropy. For comparison purposes, current co-moving anisotropy can be defined as unity [*i.e.*, $\log (t_0/t_0) = 0$].

Note that, by this definition of current co-moving anisotropy, current gravitational entropy and temperature anisotropy have maximum values, and the earliest universe has the lowest entropy and temperature anisotropy values. This approach comports with the second law of thermodynamics and the theoretical basis of the Sachs-Wolfe effect, gravitational entropy as defined by Roger Penrose, and Erik Verlinde's 'emergent gravity' theory.

SUMMARY AND CONCLUSIONS

The purpose of this chapter has been to show how one can use the FSC model of gravitational entropy to calculate cosmic radiation temperature anisotropy for any past cosmic time t since the Planck scale epoch. In the FSC model, cosmic radiation temperature anisotropy is defined by $A_t = (t/t_0)$. The derived past anisotropy value can be compared to current co-moving anisotropy defined as unity (t_0/t_0), with a log value of zero. Calculated in this way, current gravitational entropy and temperature anisotropy have maximum values, and the earliest universe has the lowest entropy and the most negative temperature anisotropy log values. This approach comports with the second law of thermodynamics and the theoretical basis of the Sachs-Wolfe effect, gravitational entropy as defined by Roger Penrose, and Erik Verlinde's 'emergent gravity' theory.

REFERENCES

[1] Tatum, E.T. (2018). How the CMB Anisotropy Pattern Could Be a Map of Gravitational Entropy. Journal of Modern Physics, 9: 1484–1490. https://doi.org/10.4236/jmp.2018.98092

[2] Sachs, R.K. and Wolfe, A.M. (1967). Perturbations of a Cosmological Model and Angular Variations of the Microwave Background. Astrophysical Journal, 147: 73. https://doi:10.1086/148982

[3] Bucher, M. (2015). Physics of the Cosmic Microwave Background Anisotropy. arXiv:1501.04288v1 [astro-ph.CO]. https://doi.org/10.1142/S0218271815300049

[4] de Bernardis, P., et al. (2000). A Flat Universe from High-Resolution Maps of the Cosmic Microwave Background Radiation. arXiv:astro-ph/0004404v1. https://doi.org/10.1038/35010035

[5] Wright, E.L., et al. (1996). Angular Power Spectrum of the Cosmic Microwave Background Anisotropy Seen by the COBE DMR. Astrophysical Journal, 464: L21–L24. https://doi.org/10.1086/310073

[6] Bennett, C.L. (2013). Nine-Year Wilkinson Microwave Anisotropy Probe (WMAP) Observations: Final Maps and Results. arXiv:1212.5225v3 [astro-ph.CO]. doi:10.1088/0067-0049/208/2/20

[7] Planck Collaboration. (2014). Planck 2013 Results. XXIII. Isotropy and Statistics of the CMB. Astronomy & Astrophysics, A23: 1–48. doi: 10.1051/0004-6361/201321534

[8] Tatum, E.T. and Seshavatharam, U.V.S. (2018). Clues to the Fundamental Nature of Gravity, Dark Energy and Dark Matter. Journal of Modern Physics, 9: 1469–1483. https://doi.org/10.4236/jmp.2018.98091

[9] Verlinde, E. (2010). On the Origin of Gravity and the Laws of Newton. arXiv:1001.0785v1 [hep-th].

[10] Verlinde, E. (2016). Emergent Gravity and the Dark Universe. aeXiv:1611.02269v2 [hep-th].

[11] Penrose, Roger. *Fashion, Faith, and Fantasy in the New Physics of the Universe*. Princeton: Princeton University Press, 2016.

[12] Penrose, Roger. (1979). Singularities and Time-Asymmetry. In Stephen W. Hawking and W. Israel (eds.). *General Relativity: An Einstein Centenary Survey*. (pp. 581–638). Cambridge: Cambridge University Press.

[13] Tatum, E.T. (2018). Flat Space Cosmology as a Model of Penrose's Weyl Curvature Hypothesis and Gravitational Entropy. Journal of Modern Physics, 9: 1935–1940. https://doi.org/10.4236/jmp.2018.910121

[14] Tatum, E.T. and Seshavatharam, U.V.S. (2018). Temperature Scaling in Flat Space Cosmology in Comparison to Standard Cosmology. Journal of Modern Physics, 9: 1404–1414. https://doi.org/10.4236/jmp.2018.97085

[15] Tatum, E.T. (2018). Why Flat Space Cosmology is Superior to Standard Inflationary Cosmology. Journal of Modern Physics, 9: 1867–1882. https://doi.org/10.4236/jmp.2018.910118

CHAPTER 10

Cosmic Time as an Emergent Property of Cosmic Thermodynamics

Abstract: This chapter, in conjunction with recent FSC publications, provides theoretical support for cosmic time being an emergent property of cosmic entropy and temperature. Therefore, if Verlinde's 'emergent gravity' theory is correct, *both time and gravity are most fundamentally emergent properties of cosmic thermodynamics.* Since emergent properties within complex systems with a huge number of degrees of freedom are often quite difficult or impossible to identify at the smallest scales, these results suggest that quantum time and quantum gravity may be no more definable than consciousness within two connected neurons. String theorists now struggling to define quantum space-time and quantum gravity should bear this in mind.*

Keywords: Flat Space Cosmology; Cosmology Theory; Emergent Gravity; Dark Matter; Cosmic Entropy; Entropic Arrow of Time; Universal Temperature; Black Holes; Mach's Principle

INTRODUCTION AND BACKGROUND

A common dictionary definition of time is that it is 'the measure of duration.' However, this definition is somewhat unsatisfying because 'duration' is simply a synonym for 'time.' Einstein's definition of time as 'what a clock measures' is correct, of course, but gets us no closer to a

*Originally published on September 4, 2018 in Journal of Modern Physics (see Appendix refs).

fundamental understanding of time. The difficulty with the philosophical question 'What is time?' is that one cannot define time in a fundamental way using only words, because any word definition of time invariably must use another word which can only be defined in terms of time [1].

The only alternative to defining time in words is to use a more rigorous form of symbolic logic, namely mathematics. Mathematics is essentially rigorous logic without the use of words, with the only exception to rigor being that some beginning set of assumptions ultimately derived from word logic is necessary as a starting point of any mathematical derivation. As shown in Gödel's incompleteness theorems, all mathematical systems must start with at least one unprovable assumption.

Aside from issues concerning starting assumptions, even with the use of mathematical equations it is not necessarily easy to define time in a fundamental way. Every student of elementary physics learns the equation $vt = s$, where v stands for velocity, t stands for time, and s stands for distance travelled. Algebraic rearrangement defines time by $t = s/v$ (*i.e.,* time as distance divided by velocity). However, this equation gets us no closer to a fundamental meaning of time than a word definition because the physics definition of velocity can only be given in terms of time. When we search all other Galilean and Newtonian physics equations incorporating a time symbol t, we invariably find at least one other variable within each equation which can only be defined in terms of time.

It was not until the 19th and early 20th centuries that time could be redefined in a non-Newtonian way. The first important breakthrough appears to have been Maxwell's discovery of a fundamental velocity (*i.e.,* speed of light c) which was entirely derivable from Faraday's laws of electromagnetism, *which did not incorporate a time variable!* Then, in 1905, Einstein conclusively proved that c is a fundamental constant of nature which is completely unshackled from the Newtonian concept of absolute time, and the tautological time definitions that come with it.

Of even greater importance, for the purposes of this introduction, was Ludwig Boltzmann's concept of entropy, which also did not incorporate a time variable. Entropy allowed for a probabilistic, but inevitable, sequence progression (*i.e., progressive change*) within highly complex systems with many degrees of freedom, including the cosmos itself. Unfortunately, Boltzmann didn't live long enough to see the potential cosmic consequences of his second law definition, because the universe in his day was widely believed to be infinite, eternal and unchanging.

It was not until Edwin Hubble's discovery of an expanding universe that Einstein and the rest of the scientific world recognized the importance of understanding universal parameters in terms of their fundamental relationship to cosmic time. This opened the way for thinking of cosmic time as being somehow deeply connected to cosmic entropy. The idea of a cosmic 'entropic arrow of time' was seriously entertained, although, until recent developments, no one had any idea how to mathematically define cosmic entropy in terms of cosmic time, particularly for infinite universe theories or cosmic models with no definable finite horizon.

The recent developments began with the Bekenstein-Hawking definition of black hole entropy [2][3] and its possible application to cosmological models according to

$$S_t \cong \frac{\pi R_t^2}{L_p^2} \tag{1}$$

wherein S_t represents cosmic entropy at time t, R_t represents the cosmic radius at time t, and L_p represents the Planck length. Furthermore, the work of Hawking and Penrose, and its implication that a universe smoothly expanding from a singularity could be modeled as a time-reversed black hole was another development. Such a model implies an ever-expanding, but definable and finite, cosmic horizon with a surface area which is directly proportional to the total entropy of the cosmic system at each point in cosmic time. Finally, the Flat Space Cosmology (FSC) model, incorporated into the spatially-flat universe Friedmann equations of Chapter five [4], completed this development. The purpose of the current chapter is to show how algebraic rearrangements from Chapter five may provide for a more fundamental understanding of cosmic time.

RESULTS

An entropic arrow of cosmic time is rigorously defined in FSC according to the following equation

$$t \cong \left(\frac{L_p}{c\sqrt{\pi}} \right) \sqrt{S_t} \tag{2}$$

wherein t represents cosmic time, L_p is the Planck length, c is speed of light and $\sqrt{S}$ is the square root of Bekenstein-Hawking's entropy S at time t [see equation (1)]. As detailed in the FSC reference, Bekenstein-Hawking's entropy is a unitless ratio, and the correct-scaling entropy term in FSC is $\sqrt{S}$. The reason for this is simply that cosmic entropy in terms of $\sqrt{S}$ scales in exactly the same way as cosmic time t (60.63 logs of 10 from the Planck scale). Furthermore, the FSC 'Universal Temperature' T_U scale [5] (see Chapter three), which is defined in a one-to-one correspondence to the Kelvin scale T by $T_U = T^2$, scales *downward* from the Planck scale temperature by the same amount (60.63 logs of 10) as time scales *upward* from the Planck scale time. This allows for a *thermodynamic arrow of time in the form of*

$$t \cong \left(\frac{\hbar L_p c^4}{32\pi^2 k_B^2 G} \right) T_U^{-1} \cong \left(\frac{\hbar L_p c^4}{32\pi^2 k_B^2 G} \right) T^{-2} \tag{3}$$

wherein T_U and T are defined as above and the other terms are well-known constants.

DISCUSSION

The above FSC definitions of cosmic time are in terms of cosmic entropy $\sqrt{S}$, cosmic Universal Temperature T_U, and temperature T in the Kelvin scale. Thus, in this cosmological model, *cosmic time appears to be fundamentally an emergent property of cosmic thermodynamics.* Furthermore, Erik Verlinde has recently suggested very persuasively that gravity and its manifestations (including dark matter and dark energy) are also emergent properties of cosmic entropy. Chapter five shows how FSC appears to be the cosmological model correlate to Verlinde's 'emergent gravity' theory. Furthermore, as detailed in the July 2018 *Journal of Modern Physics* paper entitled 'A Potentially Useful Dark Matter Index' [6], there now appear to be at least four recent observational studies [7][8][9][10] in support of Verlinde's theory [11][12], particularly with respect to observations currently attributed to dark matter. In addition, our own July 2018 *Journal of Modern Physics* paper entitled 'Equivalence Between a Gravity Field and an Unruh Acceleration Temperature Field as a Possible Clue to Dark Matter' [13] provides further theoretical support for dark matter not actually being particulate in nature. Thus, it appears likely that additional persuasive evidence in support of Verlinde's theory will be forthcoming. So, if Verlinde's theory is correct, *both time and gravity are most fundamentally emergent properties of cosmic thermodynamics.*

SUMMARY AND CONCLUSIONS

This chapter provides theoretical support for cosmic time being an emergent property of cosmic entropy and temperature. Therefore, if Verlinde's 'emergent gravity' theory is correct, *both time and gravity would appear to be most fundamentally emergent properties of cosmic thermodynamics.* Since emergent properties within complex systems with a huge number of degrees of freedom are often quite difficult or impossible to identify at the smallest scales, quantum time and quantum gravity may be no more definable than consciousness within two connected neurons. String theorists now struggling mightily to define quantum space-time and quantum gravity should bear this in mind.

REFERENCES

[1] Muller, Richard. *Now: The Physics of Time*. New York: W.W. Norton & Co., 2016.

[2] Bekenstein, J.D. (1974). Generalized Second Law of Thermodynamics in Black Hole Physics. Phys. Rev. D, 9: 3292–3300. doi:10.1103/PhysRevD.9.3292

[3] Hawking, S. (1976). Black Holes and Thermodynamics. Physical Review D, 13(2): 191–197.

[4] Tatum, E.T. and Seshavatharam, U.V.S. (2018). Clues to the Fundamental Nature of Gravity, Dark Energy, and Dark Matter. Journal of Modern Physics, 9: 1469–1483. https://doi.org/10.4236/jmp.2018.98091

[5] Tatum, E.T. and Seshavatharam, U.V.S. (2018). Temperature Scaling in Flat Space Cosmology in Comparison to Standard Cosmology. Journal of Modern Physics, 9: 1404–1414. https://doi.org/10.4236/jmp.2018.97085

[6] Tatum, E.T. (2018). A Potentially Useful Dark Matter Index. Journal of Modern Physics, 9: 1564–1567. https://doi.org/10.4236/jmp.2018.98097

[7] van Dokkum, P., et al. (2018). A Galaxy Lacking Dark Matter. Nature, 555: 629–632. doi:10.1038/nature25767

[8] Posti, L. and Helmi, A. (2018). Mass and Shape of the Milky Way's Dark Matter Halo with Globular Clusters from Gaia and Hubble. arXiv:1805.01408v1 [astro-ph.GA].

[9] Brouwer, M.M., et al. (2016). First Test of Verlinde's Theory of Emergent Gravity Using Weak Gravitational Lensing Measurements. Monthly Notices of the Royal Astronomical Society, 000: 1–14. arXiv:1612.03034v2 [astro-ph.CO].

[10] Genzel, R., et al. (2017). Strongly baryon-dominated disk galaxies at the peak of galaxy formation ten billion years ago. Nature, 543: 397–401.

[11] Verlinde, E. (2010). On the Origin of Gravity and the Laws of Newton. arXiv:1001.0785v1 [hep-th].

[12] Verlinde, E. (2016). Emergent Gravity and the Dark Universe. aeXiv:1611.02269v2 [hep-th].

[13] Tatum, E.T. and Seshavatharam, U.V.S. (2018). Equivalence Between a Gravity Field and an Unruh Acceleration Temperature Field as a Possible Clue to 'Dark Matter.' Journal of Modern Physics, 9: 1568–1572. https://doi.org/10.4236/jmp.2018.98098

CHAPTER 11

Flat Space Cosmology as a Model of Light Speed Cosmic Expansion - Implications for the Vacuum Energy Density

Abstract: Cosmologists have long ignored a stipulation by quantum field theorists that the vacuum pressure p corresponding to the zero-state vacuum energy must always be equal in magnitude to the vacuum energy density ρ (*i.e.*, $p = \rho$). Although general relativity stipulates the additional condition of proportionality between the vacuum gravitational field and $(\rho + 3p)$, the equation of state for the cosmic vacuum must fulfill both relativistic and quantum stipulations. This chapter shows how the FSC assumptions can be integrated into the Friedmann equations which include a cosmological term and a global curvature term k set to zero. The result has interesting implications for the nature of dark energy, cosmic entropy and the entropic arrow of time. The FSC vacuum energy density is shown to be equal to the cosmic fluid bulk modulus at all times, thus meeting the quantum theory stipulation of $(p = \rho)$. To date, FSC is the only viable dark energy cosmological model which has fully-integrated general relativity and quantum features, and matches current observations of an equation of state term w value of -1.0 within the margin of observational error.*

Keywords: Cosmology Theory; General Relativity; Dark Energy; Cosmic Flatness; Cosmic Entropy; Entropic Arrow of Time; Cosmic Inflation; Milne Universe; Black Holes; Cosmological Constant Problem

*Originally published on September 13, 2018 in Journal of Modern Physics (see Appendix refs).

INTRODUCTION AND BACKGROUND

Flat Space Cosmology (FSC) is a mathematical model of universal expansion which has proven to be remarkably accurate in comparison to observations [1][2][3][4]. FSC began as a heuristic model following the Penrose-Hawking implication of treating the expanding universe as a time-reversed giant black hole, one which is smoothly expanding as opposed to smoothly collapsing [5][6].

One of the results of integrating FSC into the Friedmann equations is that the following relation holds true in FSC, wherein the global curvature term k is perpetually set to zero:

$$\frac{3H^2c^2}{8\pi G} \cong \frac{\Lambda c^4}{8\pi G} \tag{1}$$

This is merely a reflection that global space-time in FSC is flat during the cosmic expansion. As stipulated by the space-time curvature rules of general relativity, a globally flat universe *must* have a net energy density of zero. Otherwise, if the positive energy density and negative energy density cosmological terms were not equal in magnitude, there would be an observable global space-time curvature representative of the greater energy density term.

Astronomical observations [7][8][9], in the context of general relativity, indicate that a mysterious energy presumably within the cosmic vacuum must be exerting a force in opposition to that of attractive gravity. Thus, this vacuum 'dark energy' is defined as a negative energy with respect to the positive energy of cosmic matter. General relativity stipulates the associated vacuum gravitational field to be proportional to $(\rho + 3p)$. Quantum field theory makes the additional stipulation that the vacuum pressure p corresponding to the zero-state vacuum energy must always be equal in magnitude to the vacuum energy density ρ (*i.e.*, $p = \rho$). Cosmologists, who seem to be particularly focused on $(\rho + 3p)$, appear not to be strictly adhering to the quantum theory stipulation. If need be, they appear willing to consider an equation of state w value other than exactly minus one. Nevertheless, the cosmic vacuum equation of state must follow *both* relativistic and quantum stipulations.

The purpose of this chapter is to show how the FSC Friedmann equations evolve further from equation (1) and what they imply, especially with respect to the vacuum energy conditions stipulated by general relativity and quantum field theory. The five basic assumptions of FSC and their observational correlations can be reviewed in Chapter one.

FLAT SPACE COSMOLOGY FRIEDMANN EQUATIONS

With respect to the Friedmann equations, those incorporating a non-zero cosmological term (*i.e.,* a dark energy term) are now the most relevant since the 1998 Type Ia supernovae discoveries. Therefore, accepting Friedmann's starting assumptions of homogeneity, isotropism and an expanding cosmic system with a stress-energy tensor of a perfect fluid, we have his cosmological equation

$$\frac{\dot{a}^2 + kc^2}{a^2} \cong \frac{8\pi G\rho + \Lambda c^2}{3} \tag{2}$$

This equation is derived from the 00 component of the Einstein field equations. Since the global curvature term k is always zero in FSC, equation (2) reduces to

$$\left(\frac{\dot{a}}{a}\right)^2 \cong H^2 \cong \frac{8\pi G\rho}{3} + \frac{\Lambda c^2}{3} \tag{3}$$

With rearrangement, we have

$$\frac{3H^2}{8\pi G} - \frac{\Lambda c^2}{8\pi G} \cong \rho \tag{4}$$

This is the relevant Friedmann equation for cosmic mass density. Multiplying all terms by c^2 gives us the relevant Friedmann equation for cosmic energy density

$$\frac{3H^2c^2}{8\pi G} - \frac{\Lambda c^4}{8\pi G} \cong \rho c^2 \tag{5}$$

At this point it is crucial to remember that Friedmann's energy density derivation of Einstein's field equations for the cosmic system as a whole (*i.e.*, globally) can be interpreted in the form of additive space-time curvatures represented by the individual terms. The first term can be read as the positive energy density (*i.e.*, the positive space-time curvature) term; the second term can be read as the negative energy density (*i.e.*, the negative space-time curvature) term; and the third term can be read as the summation (*i.e.*, *net*) energy density term for global cosmic space-time curvature. Since global space-time is treated as constantly and perfectly flat in FSC, the third term must always have a net value of zero energy density. This is entirely in keeping with the general theory of relativity, as applied to cosmology, as well as current cosmological observations of flatness (*i.e.*, critical density). Hence, in FSC

$$\frac{3H^2}{8\pi G} \cong \frac{\Lambda c^2}{8\pi G} \tag{6}$$

and

$$\frac{3H^2c^2}{8\pi G} \cong \frac{\Lambda c^4}{8\pi G} \tag{7}$$

From these respective critical mass density and energy density equations, it is obvious that the FSC model defines the Lambda term Λ by

$$\Lambda \cong \frac{3H^2}{c^2} \tag{8}$$

In FSC and other realistic linear Milne-type models, Hubble parameter H is a quantity which scales with cosmic time and is defined as

$$H \cong \frac{c}{R} \tag{9}$$

where c is the speed of light and R is the cosmic radius as defined by the Schwarzschild formula

$$R \cong \frac{2GM}{c^2} \tag{10}$$

where M represents the total matter mass of the cosmic system and G is the universal gravitational constant. Therefore, equation (8) substituted by equation (9) gives

$$\Lambda \cong \frac{3}{R^2} \tag{11}$$

So, the Lambda term Λ is also a scalar quantity (*i.e.,* like the Hubble parameter, not actually a constant) over the great span of cosmic time in the FSC model. This indicates that *FSC is a dynamic dark energy quintessence model.*

Crucially, equation (11) allows one to compare the Lambda term Λ with total entropy for the FSC cosmic system over the span of cosmic time. Recalling the Bekenstein-Hawking derivation of black hole entropy [10] [11] as directly proportional to the event horizon surface area $4\pi R^2$, we can apply their formula for cosmic entropy

$$S \cong \frac{\pi R^2}{L_p^2} \tag{12}$$

Then substituting equation (11) into equation (12) and rearranging terms

$$\Lambda \cong \frac{3\pi}{S L_p^2} \tag{13}$$

Thus, the Lambda term Λ in FSC is inversely proportional to total cosmic entropy S at all times. Substituting equation (13) into equation (8) gives

$$S \cong \frac{\pi c^2}{H^2 L_p^2} \tag{14}$$

and

$$H \cong \frac{c}{L_p}\sqrt{\frac{\pi}{S}} \tag{15}$$

And, since the reciprocal of the Hubble parameter is the measure of cosmic time t in FSC

$$t \cong \frac{L_p}{c}\sqrt{\frac{S}{\pi}} \tag{16}$$

So, cosmic time is always directly proportional to $\sqrt{S}$, with entropy S as defined by Bekenstein and Hawking. Thus, the "entropic arrow of time" is clearly defined in the FSC model.

The dark energy density cosmological term is not only expressed as $(\Lambda c^4/8\pi G)$ in FSC Friedmann equation (7), but by incorporating equation (13) into this term, we now have a dark energy density equation

$$\frac{\Lambda c^4}{8\pi G} \cong \frac{3c^4}{8GSL_p^2} \cong \frac{3H^2c^2}{8\pi G} \tag{17}$$

wherein any of these terms can be used interchangeably to quantify the absolute magnitude of the cosmic dark energy density at all times.

Since the FSC cosmic expansion model follows the Friedmann starting assumptions of homogeneity, isotropism, and an expanding cosmic system with a stress-energy tensor of an ideal fluid, one can consider the bulk modulus B of classical mechanics to apply to such a fluid. Thus, the wave velocity v_w of the cosmic fluid should have the following relationship

$$v_w = \sqrt{\frac{B}{d}} \tag{18}$$

In a realistic cosmic model, where $v_w = c$, and $d = \rho$, and B is the cosmic vacuum bulk modulus, this can be expressed as

$$c = \sqrt{\frac{B}{\rho}} \tag{19}$$

Thus,

$$B = \rho c^2 \tag{20}$$

Therefore, the pressure of this particular ideal fluid (the cosmic vacuum) must always equal its energy density by virtue of the fact that the cosmic vacuum always has a wave velocity of c. This satisfies the quantum theory ($p = \rho$) stipulation for the zero-state vacuum energy. For this reason, in the FSC cosmology model the equation of state term w value is perpetually -1.0.

DISCUSSION

Cosmologists have long ignored a stipulation by quantum field theorists that the vacuum pressure p corresponding to the zero-state vacuum energy must always be equal in magnitude to the vacuum energy density ρ (*i.e.*, $p = \rho$). While general relativity further stipulates proportionality between the repulsive gravity field of dark energy and ($\rho + 3p$), there appears to be little interest within the standard inflationary cosmology community concerning strict adherence to the above quantum theory stipulation.

The discovery of dark energy within the cosmic vacuum applies to cosmic models an accelerator pedal in opposition to the brake pedal of attractive gravity. An important question yet to be resolved is whether the cosmic acceleration value is precisely zero or *very slightly* positive.

Deep statistical analysis of the Supernova Cosmology Project compilation data shows only two remaining viable categories of dark energy cosmological models: realistic Milne-type $R_h = ct$ models (with a zero acceleration value by definition); and the standard model (with an exceedingly small positive acceleration value). A number of recent papers [12][13][14][15] show strong statistical support for the continued viability of realistic Milne-type $R_h = ct$ models.

The FSC model is the most realistic Milne-type $R_h = ct$ model to date, by virtue of the fact that it is now fully integrated into the Friedmann equations (as shown in this chapter) and shows tight correlations with the 2015 Planck Collaboration report findings, including their consensus *observational* Hubble parameter value of 67.8 +/− 0.9 km.s^{-1}.Mpc^{-1}. A realistic Milne-type model, in sharp contrast to Milne's original 'empty universe' model [16], is one which contains gravitational matter. As further evidence of the continued viability of realistic $R_h = ct$ models, Figure 5 of the following link from the Supernova Cosmology Project (SCP) is offered as proof: https://dx.doi.org/10.1088/0004-637x/746/1/85 [17]. The FSC realistic Milne-type $R_h = ct$ model is mathematically a perpetual critical density (*i.e.*, 'flat') model, as shown in the seminal FSC papers [Tatum, et al. (2015)]. Therefore, it falls on the 'flat' universe line in the SCP figure link.

Those with knowledge of the observational studies of the ratio of dark matter to visible matter realize the difficulty of determining a *precise co-moving* value for this ratio at the present time. Too little is yet known about dark matter for such precision. Galactic and peri-galactic distributions of dark matter can be surprisingly variable, as evidenced by the 29 March 2018 report in *Nature* [18] of a galaxy apparently completely lacking in dark matter! Although the 2015 Planck Collaboration consensus is a *large-scale* approximate ratio of 5.47 parts dark matter to one part visible matter, this can only be a rough estimate of the actual *co-moving* ratio, particularly if this ratio varies significantly over cosmic time. A 9-to-1 actual ratio in co-moving galaxies remains a possibility, and would change the actual ratio of total matter mass-energy to dark energy to essentially unity (*i.e.*, 50% matter and 50% dark energy). The intersection zone of tightest constraints shown in the linked figure from the Suzuki publication of the Supernova Cosmology Project should then correlate with

0.5 Ω_m and 0.5 Ω_Λ. This is one of several important *testable predictions* discriminating the FSC model from the standard model.

So long as these models are in a statistical dead heat, it is reasonable to ask if one of the two models is more compatible with the quantum theory stipulation of equality between the vacuum pressure and the vacuum energy density. The answer to this question is clearly in favor of FSC, because the predicted 50/50 percentage ratio for FSC meets the quantum theory stipulation, whereas the standard model 30/70 percentage ratio does not.

The reason that the 50/50 percentage ratio of FSC implies equality between the vacuum pressure and the vacuum energy density is contained in the final four equations of this paper [equations (17) thru (20)]. The crucial equations are (17) and (20). Only in the FSC model can the vacuum energy density cosmological term ($\Lambda c^4/8\pi G$) be shown to be equal in absolute magnitude to both the total matter mass-energy density ($3H^2c^2/8\pi G$) and the vacuum pressure represented by the bulk modulus B term in equation (20).

One of the unexpected findings of integrating the FSC model into the Friedmann equations is the intriguing *discovery of a possible close relationship between dark energy and total cosmic entropy* [see equations (11) through (17)]. Chapter five explores this topic in some detail. A review of prior publications concerning this possible relationship indicates that theoretical physicist Roger Penrose explores this subject in his recent book entitled *Fashion Faith and Fantasy in the New Physics of the Universe* on pages 275–285 [19]. Perhaps most notably, Penrose uses the Bekenstein-Hawking black hole entropy formula to derive the same equation (in rearranged form, on his page 277) as equation (11) in this chapter. This gives us great confidence that resulting FSC equations (12) through (17), showing mathematical relationships between total cosmic entropy and the Hubble parameter, cosmic time and vacuum energy density, are indeed realistic and correct.

It is important to reiterate that the equation of state for the cosmic vacuum is stipulated by *both* relativistic and quantum considerations. By apparently ignoring the stipulation from quantum field theory that ($p = \rho$) must always hold truc for thc cosmic vacuum, standard cosmology practitioners appear to have reached an impasse between general relativity and quantum theory. The FSC model, on the other hand, has fully integrated these relativistic and quantum stipulations, while maintaining remarkable accuracy with respect to current observations.

SUMMARY AND CONCLUSIONS

The FSC model of cosmology was developed as a heuristic mathematical model of the Hawking-Penrose implication that an expanding universe arising from a singularity state could be modeled as a time-reversed giant black hole. This idea was an extension of Penrose's 1965 paper on the singularities of black holes and cosmology. Hawking's doctoral thesis took the idea further by implying the validity of time-reversal in the treatment of general relativity as it concerns cosmology [Hawking and Penrose (1970)]. Finally, the FSC model completes this idea by incorporating scaling black hole equations suitable for cosmology. Thus, the proven accuracy of FSC with respect to current astronomical observations does not appear to be an accident.

To overcome any potential objections that FSC does not fit within general relativity, this chapter fully integrates FSC into the Friedmann equations. *The FSC Friedmann equations suggest a tight correlation between total dark energy, total cosmic entropy and the entropic arrow of time.*

The primary purpose of this chapter has been to show the implications of the FSC Friedmann equations with respect to the vacuum energy density. Particular attention is paid to equation (1) as it relates to equations (17) thru (20). Since a perpetually flat universe model implies (by the global curvature rules of general relativity) perpetual equality of the absolute magnitudes of global positive energy density and global negative energy density, the absolute magnitude of the vacuum pressure can be equated with the magnitude of critical energy density (now about 10^{-9} J.m^{-3}). Figure 2 in Chapter 5 nicely demonstrates this equality. *Thus, there appears to be a FSC model correlation with Verlinde's 'emergent gravity' theory [20][21].* Remembering that Friedmann's assumptions included treatment of the cosmic fluid (*i.e.*, the cosmic vacuum) as a perfect fluid, classical wave velocity equation (18) would seem to be appropriate for the FSC model. Thus, the cosmic wave velocity (speed of light c) should be directly proportional to the square root of the cosmic fluid bulk modulus, and inversely proportional to the square root of the cosmic fluid density. This relationship leads to equation (20), indicating perpetual equality between the absolute magnitudes of the vacuum pressure and the vacuum energy density. So FSC *stipulates* an equation of state w term value of -1.0 in perpetuity. This follows directly from the small number of FSC assumptions incorporated into the Friedmann equations. It is not an *ad hoc* adjustment to cosmology theory, as has clearly been the case for the various theories of cosmic inflation [22][23].

In conclusion, so long as standard cosmology proponents accept, as fact, that cosmic expansion is undergoing positive acceleration, *however small,* as implied by their claim of a 30/70 percentage ratio of total cosmic matter mass-energy to dark energy, standard cosmology is not adhering strictly to the ($p = \rho$) stipulation of quantum field theory. To date, FSC is the only viable dark energy cosmological model which has integrated general relativity and quantum features, and matches current observations of an equation of state term w value of -1.0 within the margin of observational error.

REFERENCES

[1] Tatum, E.T., Seshavatharam, U.V.S. and Lakshminarayana, S. (2015). The Basics of Flat Space Cosmology. International Journal of Astronomy and Astrophysics, 5: 116–124. http://doi.org/10.4236/ijaa.2015.52015

[2] Tatum, E.T., Seshavatharam, U.V.S. and Lakshminarayana, S. (2015). Thermal Radiation Redshift in Flat Space Cosmology. Journal of Applied Physical Science International, 4(1): 18–26.

[3] Tatum, E.T., Seshavatharam, U.V.S. and Lakshminarayana, S. (2015). Flat Space Cosmology as an Alternative to LCDM Cosmology. Frontiers of Astronomy, Astrophysics and Cosmology, 1(2): 98–104. http://pubs.sciepub.com/faac/1/2/3

[4] Planck Collaboration XIII. (2015). Cosmological Parameters. http://arxiv.org/abs/1502.01589

[5] Hawking, S. and Penrose, R. (1970). The Singularities of Gravitational Collapse and Cosmology. Proc. Roy. Soc. Lond. A, 314: 529–548.

[6] Penrose, Roger. *Fashion, Faith, and Fantasy in the New Physics of the Universe*. Princeton: Princeton University Press, 2016.

[7] Perlmutter, S., et al. (1999). The Supernova Cosmology Project, Measurements of Omega and Lambda from 42 High-Redshift Supernovae. Astrophysical Journal, 517: 565–586. [DOI], [astro-ph/9812133].

[8] Schmidt, B. et al. (1998). The High-Z Supernova Search: Measuring Cosmic Deceleration and Global Curvature of the Universe Using Type Ia Supernovae. Astrophysical Journal, 507: 46–63.

[9] Riess, A.G., et al. (1998). Observational Evidence from Supernovae for an Accelerating Universe and a Cosmological Constant. Astronomical Journal, 116(3): 1009–38.

[10] Bekenstein, J.D. (1974). Generalized Second Law of Thermodynamics in Black Hole Physics. Phys. Rev. D, 9: 3292–3300. doi:10.1103/PhysRevD.9.3292

[11] Hawking, S. (1976). Black Holes and Thermodynamics. Physical Review D, 13(2): 191–197.

[12] Tutusaus, I., et al. (2017). Is Cosmic Acceleration Proven by Local Cosmological Probes? Astronomy & Astrophysics, 602_A73.arXiv:1706.05036v1 [astro-ph.CO].

[13] Nielsen, J.T., et al. (2015). Marginal Evidence for Cosmic Acceleration from Type Ia Supernovae. doi: 10.1038/srep35596. arXiv:1506.01354v1.

[14] Wei, J.J., et al. (2015). A Comparative Analysis of the Supernova Legacy Survey Sample with ΛCDM and the $R_h = ct$ Universe. Astronomical Journal, 149: 102–112.

[15] Melia, F. (2012). Fitting the Union 2.1 SN Sample with the $R_h = ct$ Universe. Astronomical Journal, 144. arXiv:1206.6289 [astro-ph.CO].

[16] Milne, E.A. (1933). Z. Astrophysik, 6: 1–35.

[17] Suzuki, et al. (2011). The Hubble Space Telescope Cluster Supernovae Survey: V. Improving the Dark Energy Constraints Above Z>1 and Building an Early-Type-Hosted Supernova Sample. arXiv.org/abs/1105.3470.

[18] Van Dokkum, P., et al. (2018). A Galaxy Lacking Dark Matter. Nature, 555: 629–632. doi: 10.1038/nature25767

[19] Penrose, Roger. (2016). *Fashion, Faith, and Fantasy in the New Physics of the Universe*. Princeton: Princeton University Press.

[20] Verlinde, E. (2010). On the Origin of Gravity and the Laws of Newton. arXiv:1001.0785v1 [hep-th].

[21] Verlinde, E. (2016). Emergent Gravity and the Dark Universe. arXiv:1611.02269v2

[22] Guth, Alan. *The Inflationary Universe: The Quest for a New Theory of Cosmic Origins*. New York: Basic Books, 1997.

[23] Steinhardt, P.J. (2011). The Inflation Debate: Is the Theory at the Heart of Modern Cosmology Deeply Flawed? Scientific American, 304(4): 18–25.

CHAPTER 12

My C.G.S.I.S.A.H. Theory of Dark Matter

Abstract: Theory and observations concerning the cosmic reionization epoch are briefly discussed in the context of recent observations attributed to dark matter. A case is made that cold ground state interstellar atomic hydrogen of average density of about one atom per cubic centimeter (1.67×10^{-21} kg.m^{-3} or 1.67×10^{-24} g.cm^{-3}) appears to be the most likely candidate to explain these observations.*

Keywords: Dark Matter; Early Universe; Reionization Epoch; Dark Age; Cosmology Observations; Galaxies:ISM; ISM:atoms; Radio Lines:ISM

INTRODUCTION AND BACKGROUND

In May of 2019, one of us (E.T.T.) shared his C.G.S.I.S.A.H. theory of dark matter [1] with colleagues at the dark matter workshop sponsored by the World Science Festival. What follows is a brief note concerning the new constraints on dark matter and a discussion of his conjecture and its observational predictions.

Convincing observational support for dark matter begins with the publication by Rubin and Ford [2] concerning unexpected galactic rotation curves. These observations, soon followed by others [3], provide strong support that an invisible (*i.e.,* 'dark') form of gravitationally attractive matter within the interstellar vacuum is contributing to galaxies approximately 5–10 times the total mass of the visible galactic matter (*i.e.,* stars,

---

*Originally published on July 5, 2019 in Journal of Modern Physics (see Appendix refs).

warm molecular gas clouds, and dust). By 'invisible' it is meant that this matter is not emitting any detectable light.

It has subsequently become apparent that one can further observe the *effect* of this dark matter by its gravitational lensing properties. By these observations, there appears to be a roughly spherical cloud (*i.e.,* a 'halo') of dark matter gas or superfluid extending up to approximately 200 kpc from the observed galactic centers. Dark matter is also nearly collisionless due to a low scattering cross-section, as deduced from Tucker's observations of the bullet cluster [4] and other colliding galaxy clusters. Furthermore, the Planck Collaboration report [5] of the cosmic microwave background (CMB) anisotropy indicates that dark matter was present at the time of the recombination/decoupling epoch. It is even postulated that dark matter has been the seeding structural scaffolding for the further formation of galaxies, galaxy clusters and filaments in the subsequent evolution of the universe.

In recent years, numerous theories and detection methods have been proposed for dark matter with the above properties. While it is not within the scope of this paper to review the many publications on this subject, three important publications in 2018 and two important publications in 2019 deserve special mention herein.

The first of these is Barkana's review [6] of the reionization epoch ('cosmic dawn') 21-cm observations. These observations constrain dark matter to a very slow-moving (*i.e.,* cold) particle with a mass-energy of no greater than 2–3 GeV. Furthermore, the graph on page 9 of the Barkana reference shows a very tight correlation between a dark matter particle of about 0.938 GeV and the minimum possible 21-cm brightness temperature T_{21} at redshift $z = 17$. Thus, *atomic hydrogen appears to be the only baryon not yet ruled out by these new tight constraints*.

The second reference of importance in 2018 is Posti and Helmi's analysis [7] of *Gaia* data extracted from a 20 kpc (65.2 thousand light-years) radius halo sphere centered at the Milky Way center. From their analysis one can deduce the ratio of dark matter to visible matter within this halo sphere to be approximately 1.37 to 0.54, or 2.54 to one. This ratio will be further addressed in the Results section to follow.

The third reference of importance in 2018 is physicist Stacy McGaugh's publication entitled, 'Strong Hydrogen Absorption at Cosmic Dawn: The Signature of a Baryonic Universe' [8]. One should carefully read the

McGaugh reference for the reasoning that *the cosmic dawn observations fit best for baryonic dark matter. Thus, nonbaryonic proposals for dark matter do not appear to be necessary.*

The first reference of importance in 2019 is the Read publication [9] which provides support for 'dark matter heating' within active galactic centers. This process may explain why active galactic centers tend to have a somewhat shallower dark matter core. Thus, dark matter heating may be an important variable in understanding its perplexing spatial distribution, particularly with respect to the dark matter 'cusp-core problem.'

The second reference of importance in 2019 is the March online report [10] of the *Gaia*-Hubble Collaboration. Here, for the first time, one can have confidence that the 'visible matter mass' of the Milky Way is approximately 250 billion solar masses ($M_\odot$). Therefore, if one can assume that this visible matter mass is roughly confined to within the 20 kpc Posti and Helmi radius halo sphere, their 2.54 ratio would imply approximately 635 billion $M_\odot$ of dark matter within 20 kpc of the Milky Way center.

With all of the above observations concerning dark matter, one can now construct the following table (**Table 1**) of these features with the relevant references listed in the right-hand column:

Given these features characteristic of dark matter, it is useful to review what observations suggest about the evolution of the universe since the recombination/decoupling epoch. During the adiabatic cooling period of the cosmic 'dark age' the positive feedback of gravitational attraction is

Table 1 Dark Matter Features and Relevant References.

Dark Matter Features	References
Cold (*i.e.,* low velocity)	Barkana
No Emissions (*i.e.,* dark)	Rubin & Ford
Collisionless (*i.e.,* low cross-section)	Tucker
Baryon Expected	McGaugh
Mass-Energy less than 3 GeV	Barkana
Dark Matter $M_{20\,kpc}$ 635 billion $M_\odot$	Gaia-Hubble/Posti & Helmi
Central DM Heating ('coring')	Read
CMB Decoupling at Dawn	Astrobaki/McGaugh
Structural Scaffold	Planck
Existence at CMB emission	Planck

thought to have accentuated the anisotropy we now observe in the cosmic microwave background (CMB) by creating centripetal movements of the atomic hydrogen within the denser regions of the CMB map. In contrast to these collapsing and swirling clouds of the nascent stars and galaxies, the intervening atomic hydrogen within the minimum density regions of the CMB map is thought to have been relatively motionless (i.e., colder). With the continuing cosmic expansion, this intervening atomic hydrogen, the primary matter in regions we now refer to as the intergalactic and interstellar vacuum, would have ultimately become so sparse as to be nearly collisionless and predominantly confined to the ground state (except where in close proximity to the nascent stars). At the beginning of the reionization epoch (*i.e.,* 'cosmic dawn') the Wouthuysen-Field effect of the Lyman-alpha radiation of the first stars should have reduced the temperature T_G of ground state interstellar atomic hydrogen to well below the CMB radiation temperature T_R [11]. Such a temporary decoupling from the CMB radiation temperature would have eventually resolved due to the increasing stellar black body radiation closing out the cosmic dawn epoch.

Astrophysical observations of the 21-cm absorption line in the redshift z range corresponding (in standard cosmology) to approximately 110–250 million years after the Big Bang show evidence of a process very much like this, as seen in **Figure 1** [12]. However, the conventional wisdom is that a mysterious nearly collisionless non-baryonic cold dark matter must have also been present in the interstellar vacuum, as a required intermediary in this process.

Unfortunately, the reasoning that such an intermediary nonbaryonic matter was required for this process is also somewhat mysterious, because a temporary decoupling from the CMB radiation temperature is to be expected in a purely baryonic universe (see McGaugh [8]).

Perhaps the ongoing search for exotic dark matter also reflects a misunderstanding about the current abundance of cold ground state interstellar atomic hydrogen in comparison to the constituents of the visible stars, warm molecular gas clouds and dust in our galaxy. It should be remembered that ground state interstellar atomic hydrogen coupled to the CMB radiation temperature (as was also undoubtedly present in great abundance during the 'dark age') is essentially invisible to modern detectors,

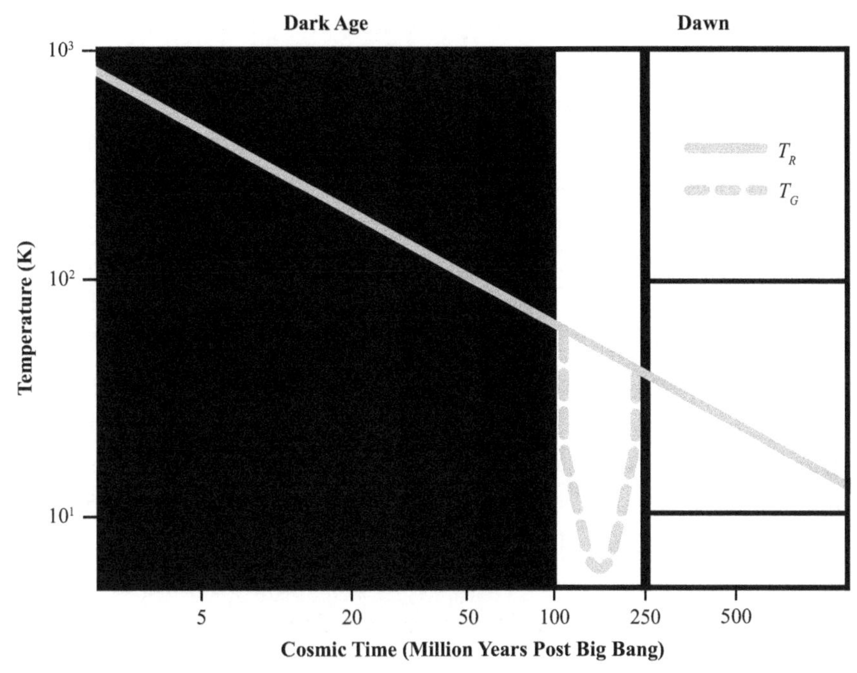

Figure 1 Radiation Temperature (T_R) and Gas Temperature (T_G) vs Time.

except where its characteristic 21-cm absorption line is 'backlit' by distant starlight.

The Milky Way disc rotates with a period of approximately 250 million years [13]. Based upon the Baryonic Tully-Fisher relation [14] and the March 2019 *Gaia*-Hubble Collaboration report, the sum total mass of the visible stars, warm molecular gas clouds and dust in the Milky Way is reliably estimated to be 250 billion $M_{\odot}$. And yet, the calculated amount of ground state interstellar atomic hydrogen coupled to the CMB radiation temperature within a 20 kpc halo radius of the Milky Way center actually dwarfs this total visible matter mass estimate (see calculation below). This reflects the vastness of the interstellar vacuum in comparison to the visible matter.

RESULTS

Line-of-sight measurements of the hyperfine 21-cm absorption line (within the light from distant stars of a known distance from the observer) allow one to estimate an average density of cold ground state interstellar atomic hydrogen of approximately one atom/cm^3 (or 1.67×10^{-21} $kg.m^{-3}$) [15] [16] [17].

One can now use this knowledge in the context of Posti and Helmi's recent Gaia survey analysis of the Milky Way (see reference [7]). They report the ratio of dark matter to visible matter within a 20 kpc spherical halo radius of the galactic center to be approximately 1.37 to 0.54. This simplifies to a ratio of approximately 2.54. If we assume the above current best estimate of the Milky Way visible matter mass (250 billion $M_\odot$ is equal to 4.97×10^{41} kg) and divide that by the volume of a galactic halo sphere of 20 kpc radius (9.85×10^{62} m^3), the average visible matter density within that galactic halo sphere is 5.05×10^{-22} $kg.m^{-3}$, approximately one-third of the above-mentioned average density of cold ground state interstellar atomic hydrogen! Multiplying 1.67×10^{-21} $kg.m^{-3}$ by a 0.83 correction factor (for the expected slightly lower ground state atomic hydrogen density in the halo sphere portion outside the galactic disk) times the 20 kpc radius galactic halo sphere volume gives an estimated mass of cold ground state interstellar atomic hydrogen of 1.37×10^{42} kg, in other words approximately 689 billion $M_\odot$, within that sphere. The corresponding 2.76 ratio (from dividing 689 billion $M_\odot$ by 250 billion $M_\odot$) is well within the margin of error of Posti and Helmi's observed ratio of dark matter to visible matter for the same 20 kpc radius galactic halo sphere.

DISCUSSION

The calculations made in the Results section suggest the strong possibility that cold ground state interstellar atomic hydrogen averaging approximately one atom/cm^3 is what we currently refer to as cold dark matter (CDM). The following table (**Table 2**) compares the above-mentioned dark matter features with sparse interstellar atomic hydrogen coupled to the CMB temperature.

The origin of the C.G.S.I.S.A.H. acronym becomes apparent by reading down the letters in the right-hand column, which are abbreviations for the top five rows of the table. The abbreviation W-F effect stands for the Wouthuysen-Field effect on ground state neutral atomic hydrogen due to Lyman-alpha radiation beginning with the first starlight of cosmic dawn.

Table 2 Dark Matter Features vs Interstellar H Features.

Dark Matter Features	Interstellar H at 1 Atom/cm^3	CDM
Cold (*i.e.*, low velocity)	CMB Equilibrated	C
No Emissions (*i.e.*, dark)	Ground State	GS
Collisionless (*i.e.*, low cross-section)	Interstellar/Sparse	IS
Baryon Expected	Atomic H	A
Mass-Energy less than 3 GeV	0.938 GeV For Neutral H	H
Dark Matter $M_{20\,kpc}$ 635 billion $M_{\odot}$	689 billion $M_{\odot}$	
Central DM Heating ('coring')	Loses Ground State	
CMB Decoupling at Dawn	W-F Effect	
Structural Scaffold	Most Abundant Atom	
Existence at CMB emission	Most Abundant Atom	

There is a nice discussion of this temporary CMB decoupling phenomenon in the AstroBaki reference ([11]). According to this reference, '... the W-F effect remains the dominant effect until reionization is complete.' Once reionization was complete, the interstellar atomic hydrogen presumably became once again coupled to the CMB temperature, which is assumed to be the case at present.

As for future observable consequences of my dark matter conjecture presented herein, one can point to the ongoing refinement of observational constraints on the mass-energy of the dark matter particle. The studies to date appear to eliminate any baryonic particle much greater than about 1 GeV (see Barkana [6]). However, they do not yet exclude neutral atomic hydrogen, with its mass-energy of 0.938 GeV. I predict that these constraints will further tighten around a dark matter particle with a mass-energy of 0.938 GeV. Furthermore, the sophisticated dark matter/baryon interaction simulations being conducted at the Kavli Institute for Particle Astrophysics and Cosmology have not yet simulated the dark matter candidate in these interactions as cold ground state interstellar atomic hydrogen of average density of about 1.67×10^{-21} kg.m^{-3} (1.67×10^{-24} g.cm^{-3}) [R. Wechsler, Director, per verbal communication with this author (E.T.T.) on May 30, 2019]. It is predicted that such simulations will correlate nicely with dark matter observations, even to the extent of simulating central galactic coring (*i.e.*, relative dark matter depletion) due to 'dark matter heating' within active galactic centers (see Read [9]). Thus, the previously unexplained galactic and peri-galactic dark matter spatial distribution may be best understood in terms of heating and other dynamic effects upon the distribution of cold ground state interstellar atomic hydrogen.

SUMMARY AND CONCLUSIONS

For the above theoretical and observational considerations, the distinct possibility that the dark matter candidate could ultimately prove to be the ubiquitous but incredibly sparse (and thus nearly collisionless) cold ground state interstellar atomic hydrogen must be seriously entertained.

REFERENCES

[1] Tatum, E.T. (2019). My CGSISAH Theory of Dark Matter. Journal of Modern Physics. 10: 881–887. https://doi.org/10.4236/jmp.2019.108058

[2] Rubin, V. and Ford, W.K. (1970). Rotation of the Andromeda Nebula from a Spectroscopic Survey of Emission Regions. The Astrophysical Journal. 159, 379 (doi:10.1086/150317)

[3] Rubin, V., et al. (1985). Rotation Velocities of 16 SA Galaxies and a Comparison of Sa, Sb, and SC Rotation Properties. The Astrophysical Journal. 289, 81 (doi:10.1086/162866)

[4] Tucker, W. (2006). Recent and Future Observations in the X-ray and Gamma-ray Bands: Chandra, Suzaku, GLAST, and NuSTAR, AIP Conference Proceedings. 801, 21 (arXiv:astro-ph/0512012 doi:10.1063/1.2141828)

[5] Aghanim, N., et al. (2018). Planck 2018 Results VI. Cosmological Parameters. http://arXiv:1807.06209v1

[6] Barkana, R. (2018). Cosmic Dawn as a Dark Matter Detector (arXiv:1803.06698v1) p. 9.

[7] Posti, L. and Helmi, A. (2018). Mass and Shape of the Milky Way's Dark Matter Halo with Globular Clusters from Gaia and Hubble (arXiv:1805.01408v1)

[8] McGaugh, S.S. (2018). Strong Hydrogen Absorption at Cosmic Dawn: The Signature of a Baryonic Universe (arXiv:1803.02365v1)

[9] Read, J.I. (2019). Dark Matter Heats Up in Dwarf Galaxies (arXiv:1808.06634v2)

[10] Gaia-Hubble Collaboration. (2019). http://sci.esa.int/hubble/61198-hubble-and-gaia-accurately-weigh-the-milky-way-heic1905/

[11] AstroBaki. (2017). https://casper.ssl.berkeley.edu/astrobaki/index.php/Wouthuysen_Field_effect

[12] Bowman, J.D. (2018). An Absorption Profile Centered at 78 Megahertz in the Sky-Averaged Spectrum, Nature, 555, 67 (doi:10.1038/nature25792)

[13] Morris, M. (2001). The Milky Way, *The World Book Encyclopedia*, p. 551.

[14] Torres-Flores, S., et al. (2011). GHASP: An Hα Kinematic Survey of Spiral and Irregular Galaxies—IX. The NIR, Stellar and Baryonic Tully–Fisher Relations (arXiv:1106.0505)

[15] Pananides, N.A. and Arny, T. (1979). *Introductory Astronomy,* 2nd. Ed. Reading: Addison-Wesley Publishing, p. 293.

[16] Chaisson, E. and McMillan, S. (1993). *Astronomy Today*. New York: Prentice Hall, p. 418.

[17] Mammana, D.L. (2000). *Interstellar Space*. New York: Popular Science, p. 220.

CHAPTER 13

How the Dirac Sea Idea May Apply to a Spatially-Flat Universe Model (A Brief Review)

Abstract: The famous Dirac sea idea can be resurrected if one replaces the concept of positive and negative matter mass with positive and negative energy. Utilizing this concept, the perpetually spatially-flat matter-generating FSC model can be shown to be a realistic Milne 'empty universe' model. Furthermore, this may be why $R_h = ct$ cosmological models like FSC show an excellent statistical fit with the accumulated data of the Supernova Cosmology Project.*

Keywords: Dirac Sea; Dirac Equation; Flat Space Cosmology; Dark Energy; Dark Matter; Inflationary Cosmology; Supernova Cosmology Project; $R_h = ct$ Models

INTRODUCTION AND BACKGROUND

Mathematical physicist Paul Dirac is perhaps best known for the Dirac equation in its many forms [1]. Not only was he largely responsible for making quantum mechanics relativistic, but his 'hole theory' (based upon his equation) suggested to him something about the nature of the cosmic vacuum. Dirac believed that his equation implied that the vacuum could be a many-body quantum state in which all of the negative energy eigenstates (holes) are occupied. In terms of electron eigenstates, for instance, Dirac pictured a 'sea' of electrons occupying negative-energy

*Originally published on July 17, 2019 in Journal of Modern Physics (see Appendix refs).

electron eigenstates. Unfortunately, his 'Dirac sea' idea was interpreted to imply holes of *negative mass matter*, which is believed to be impossible. The term 'antimatter' is used not to imply negative mass matter, but rather gravitationally-attractive positive-energy matter of the same mass as its positive-energy partner, yet *opposite in quantum spin and charge*. The combined mass-energy of an electron and positron, for instance, sums to twice that of the electron alone, rather than summing to zero. Despite the initial apparent failure of the Dirac sea idea, the Dirac equation was correctly credited with predicting oppositely-charged matter of equal positive mass, for which the term antimatter is restricted.

On a parallel track, there is the currently-favored theory that our universe may have started from a zero-energy state and undergone a brief (10^{-32} s) period of 'cosmic inflation' in which all matter and antimatter were created. Nevertheless, this nearly instantaneous matter-generating universe theory would appear to violate conservation of energy. This did not escape notice by its inventor, Alan Guth [2][3]. Guth has sometimes referred to his theory as a 'free lunch' idea. Other somewhat modified inflationary theories [4][5] of a nearly instantaneous matter-generating universe have followed, although the problem of energy conservation violation appears to be inherent in all such theories [6].

Recently, there have been proposed several *perpetually* matter-generating universe theories [7][8][9][10][11][12][13][14], which smoothly expand, are not inflationary in nature, and do not appear to violate conservation of energy. They are also consistent with current observations of a spatially-flat universe. One of the most successful of these theories, in terms of predicting current observations, is the 'Flat Space Cosmology' (FSC) model. By following its five basic assumptions, the heuristic FSC model perpetually maintains the Friedmann critical mass density for a spatially-flat universe ($\rho_0 = 3H_0^2/8\pi G$) for any theoretical time of observation (0). Furthermore, the FSC model tightly correlates the current redshifted cosmic microwave background (CMB) temperature of 2.72548 K with a current *predicted* Hubble parameter value of 66.9 km.s^{-1}.Mpc^{-1}. This Hubble parameter value fits within the tight constraints of the 2018 Planck Collaboration [15] and 2018 Dark Energy Survey [16] reports.

RELEVANCE OF THE DIRAC SEA IDEA TO THE FSC MODEL

The relevance of the Dirac sea idea to the FSC model pertains to its perpetual matter generation and its perpetual spatial flatness. As detailed in several 2018 publications [11][12][13], a *globally* perpetually-flat spacetime implies that the cosmological model must always maintain equal amounts of positive and negative energy. Otherwise, the more dominant energy density component would contribute an observable curvature signifying either cosmic deceleration (positive curvature) or cosmic acceleration (negative curvature). As the excellent statistical fit between $R_h = ct$ cosmological models and observations of the Supernova Cosmology Project indicates, the expansion of our universe appears to be coasting at constant velocity [7][17][18][19][20]. All $R_h = ct$ models have this 'coasting at constant velocity' feature. For specifics concerning the basic features of $R_h = ct$ models, the reader is encouraged to start with these references. FSC is one such $R_h = ct$ model.

The 'Dirac sea' idea can be resurrected if one considers a dichotomy of positive and negative *energy* states popping into and out of existence within the vacuum. If one follows the convention that all units of matter mass-energy are 'positive' energy, then one can consider the 'holes' in the Dirac sea to be similarly-sized units of 'negative' vacuum energy. Furthermore, since the FSC model uses such a sign convention, the negative energy holes in the FSC Dirac sea can now be understood to be units of dark energy exactly offset by the units of matter mass-energy produced in the FSC vacuum. By this perpetually ongoing process, the FSC model accumulates increasingly positive (matter) energy and increasingly negative dark energy of equal magnitude, always *summing to zero total energy*. In this way, a universe which begins in a zero-energy state maintains perpetual conservation of total (*i.e.*, global) cosmic energy.

RESULTS: EVIDENCE IN SUPPORT OF FSC AND DIRAC

The 'net zero energy' FSC model can now be contrasted with standard inflationary cosmology, which considers such models to be unrealistic 'empty universe' models. The phrase 'empty universe' has generally been applied in the past to Milne-type models containing no matter. However, the current $R_h = ct$ models contain matter and, as such, are considered to be more realistic than Milne's original conception. If one looks carefully at the open-source graphs published by the Supernova Cosmology Project (SPC) [21], one can see the excellent statistical fit of the 'empty' universe line demarcating the boundary between accelerating and decelerating universal expansions. For instance, the FSC cosmic model follows the 'empty' line demarcating between deceleration and acceleration in Figure 2 of the following link: https://dx.doi.org/10.4236/jmp.2020.1110091

This 'empty' universe line falls exactly where the FSC and other $R_h = ct$ models fall. One can also readily see how it is that $R_h = ct$ universe models appear to show an excellent statistical fit with SCP observations to date.

The significance of the use of the FSC model to resurrect the Dirac sea concept (at least in terms of opposite sign cosmic *energies*) is perhaps best seen in **Figure 1**. This graph is copied, in slightly modified form, from Chapter five [22]. It incorporates the Bekenstein-Hawking entropy into the model in order to represent the cosmic clock as well as the 'entropic arrow of time.'

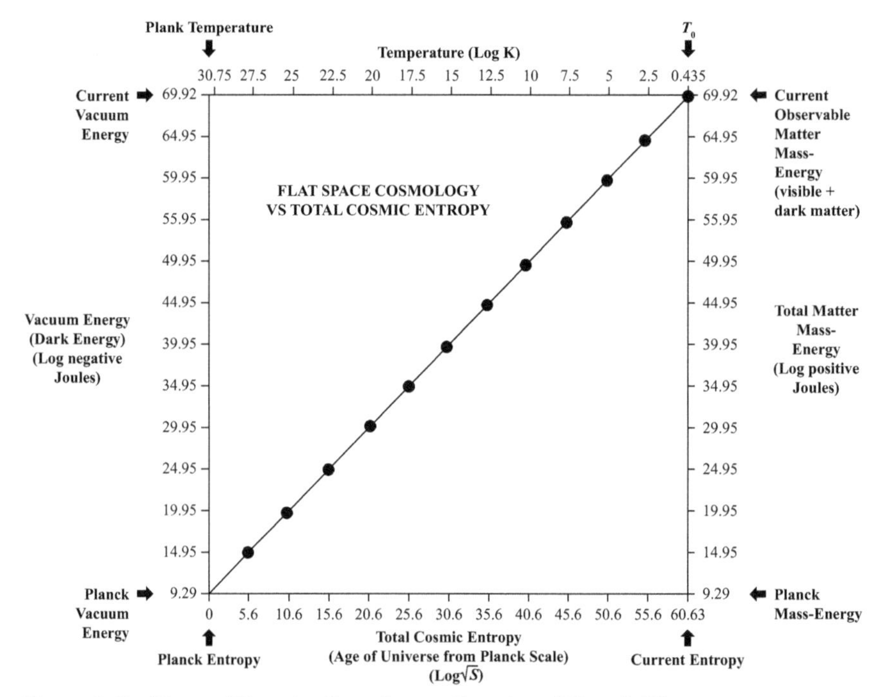

Figure 1 Positive and Negative Energies as a Function of Cosmic Time.

One can readily see that the magnitude of *positive* matter mass-energy (visible plus dark matter) of the FSC model scales in exactly the same way as the magnitude of *negative* dark energy scales. Thus, the 'net zero energy' of the universe as a global object is always maintained. In this context, the 'net zero energy' FSC model can be thought of as a realistic Milne-type 'empty universe' model!

DISCUSSION AND SUMMARY

This chapter provides a brief look at Dirac's thought process concerning how the cosmic vacuum might behave if it follows his famous equation. The 'Dirac sea' idea is resurrected in terms of a zero-point energy vacuum in which energy has positive (*i.e.,* matter) and negative (*i.e.,* dark energy) values always summing to zero (*i.e.,* 'net zero energy'). As it turns out, the FSC model, by its perpetual matter generation and its perpetual spacetime flatness, can be seen as a realistic Milne-type 'empty universe' model. The genius of Paul Dirac and his equation can once again be readily seen when his 'Dirac sea' idea for positive and negative *matter* (thought to be an impossibility) is resurrected in terms of positive and negative *energy*.

REFERENCES

[1] Wikipedia contributors. (2019, June 1). Dirac equation. In Wikipedia, The Free Encyclopedia. June 16, 2019, from https://en.wikipedia.org/wiki/Dirac_equation

[2] Guth, A.H. (1981). Inflationary Universe: A Possible Solution to the Horizon and Flatness Problems. Phys. Rev. D, 23: 347.

[3] Guth, Alan. *The Inflationary Universe: The Quest for a New Theory of Cosmic Origins*. New York: Basic Books, 1997.

[4] Albrecht, A. and Steinhardt, P.J. (1982). Cosmology for Grand Unified Theories with Radiatively-Induced Symmetry Breaking. Physical Review Letters, 48: 1220.

[5] Linde, A.D. (1982). A New Inflationary Universe Scenario: A Possible Solution of the Horizon, Flatness, Homogeneity, Isotropy, and Primordial Monopole Problems. Physics Letters, 108B: 389–92.

[6] Steinhardt, P.J. (2011). The Inflation Debate: Is the Theory at the Heart of Modern Cosmology Deeply Flawed? Scientific American, 304(4): 18–25.

[7] Melia, F. (2012). Fitting the Union 2.1 SN Sample with the Rh = ct Universe. Astronomical Journal, 144: arXiv:1206.6289 [astro-ph.CO]

[8] Tatum, E.T., Seshavatharam, U.V.S. and Lakshminarayana, S. (2015). The Basics of Flat Space Cosmology. International Journal of Astronomy and Astrophysics, 5: 116–124. http://dx.doi.org/10.4236/ijaa.2015.52015

[9] Tatum, E.T., Seshavatharam, U.V.S. and Lakshminarayana, S. (2015). Thermal Radiation Redshift in Flat Space Cosmology. Journal of Applied Physical Science International, 4(1): 18–26.

[10] Tatum, E.T., Seshavatharam, U.V.S. and Lakshminarayana, S. (2015). Flat Space Cosmology as an Alternative to LCDM Cosmology. Frontiers of Astronomy, Astrophysics and Cosmology, 1(2): 98–104. http://pubs.sciepub.com/faac/1/2/3 doi:10.12691/faac-1-2-3

[11] Tatum, E.T. (2018). Why Flat Space Cosmology Is Superior to Standard Inflationary Cosmology. Journal of Modern Physics, 9: 1867–1882. https://doi.org/10.4236/jmp.2018.910118

[12] Tatum, E.T. and Seshavatharam, U.V.S. (2018). Flat Space Cosmology as a Model of Light Speed Cosmic Expansion - Implications for the Vacuum Energy Density. Journal of Modern Physics, 9: 2008–2020. https://doi.org/10.4236/jmp.2018.910126

[13] Tatum, E.T. and Seshavatharam, U.V.S. (2018). Temperature Scaling in Flat Space Cosmology in Comparison to Standard Cosmology. Journal of Modern Physics, 9: 1404–1414. https://doi.org/10.4236/jmp.2018.97085

[14] Sapar, A. (2019). A Perpetually Mass-Generating Planckian Universe. Proceedings of the Estonian Academy of Sciences, 68(1): 1–12. https://doi.org/10.3176/proc.2019.1.01

[15] Aghanim, N., et al. (2018). Planck 2018 Results VI. Cosmological Parameters. http://arXiv:1807.06209v1

[16] Macaulay, E., et al. (2018). First Cosmological Results Using Type Ia Supernovae from the Dark Energy Survey: Measurement of the Hubble Constant. arXiv:1811.02376v1

[17] Nielsen, J.T., et al. (2015). Marginal Evidence for Cosmic Acceleration from Type Ia Supernovae. Scientific Reports, 6: Article number 35596. doi:10.1038/srep35596. arXiv:1506.01354v1.

[18] Jun-Jie Wei, et al. (2015). A Comparative Analysis of the Supernova Legacy Survey Sample with ΛCDM and the R_h = ct Universe. Astronomical Journal, 149: 102.

[19] Tutusaus, I., et al. (2017). Is Cosmic Acceleration Proven by Local Cosmological Probes? Astronomy & Astrophysics, 602_A73. arXiv:1706.05036v1 [astro-ph.CO].

[20] Dam, L.H., et al. (2017). Apparent Cosmic Acceleration from Type Ia Supernovae. Mon. Not. Roy. Astron. Soc. arXiv:1706.07236v2 [astro-ph.CO]

[21] Perlmutter, S. (2016). Supernova Cosmology Project. Berkeley (US): Lawrence Berkeley National Laboratory; [accessed 2018 June 16]. http://www.supernova.lbl.gov

[22] Tatum, E.T. and Seshavatharam, U.V.S. (2018). Clues to the Fundamental Nature of Gravity, Dark Energy and Dark Matter. Journal of Modern Physics, 9: 1469–1483. https://doi.org/10.4236/jmp.2018.98091

CHAPTER 14

A Universe Comprised of 50% Matter Mass-Energy and 50% Dark Energy

Abstract: The new C.G.S.I.S.A.H. theory of dark matter (see Chapters 12 and 16) is used to appropriately classify and quantitate the previously-ignored cold ground state neutral atomic hydrogen within the intergalactic vacuum. A surprising discovery is demonstrated in the Results section that approximately one-fifth of the cosmic critical density can be attributable to intergalactic cold ground state neutral atomic hydrogen, despite the fact that it amounts to only about one atom per cubic meter of the intergalactic vacuum! By subtracting this quantity of the critical density from the dark energy ledger column and adding it to the total matter mass-energy ledger column, our current universe appears to be equally proportioned between total matter mass-energy and dark energy. This has been a longstanding prediction of the Flat Space Cosmology model.*

Keywords: Dark Matter; Dark Energy; CGSISAH Theory; Flat Space Cosmology; Dirac Sea; Intergalactic Medium; Interstellar Medium; ΛCDM Concordance Model

INTRODUCTION AND BACKGROUND

Before 1998, the energy density of the universe was believed to be wholly comprised of approximately 5% visible matter (*i.e.*, the visible stars, warm

---

*Originally published on August 14, 2019 in Journal of Modern Physics (see Appendix refs).

molecular gas clouds and cosmic dust), an unknown quantity of dark matter within and haloed around the visible galaxies, and a comparatively negligible amount of radiation energy. Observations prior to 1998 consistent with current spatial flatness provided support for the idea that our universe is at, or very near, critical density. And yet there didn't seem, at the time, to be sufficient galactic and peri-galactic (*i.e.,* virial) dark matter to account for this. However, since the 1998 discovery of dark energy, the balance of the previously missing critical density has been wholly ascribed to a mysterious energy within the intergalactic vacuum. As such, cosmological general relativity equations now incorporate a 'cosmological constant' to represent this non-matter dark vacuum energy.

Furthermore, observations of galactic rotation [1], the Milky Way [2], and the cosmic microwave background (CMB) anisotropy appear to confirm that the galactic and peri-galactic dark matter mass is approximately 5 to 5½ times the visible galactic matter mass. For instance, the *Gaia*-Hubble Collaboration reported in March 2019 that the Milky Way has a total virial mass of approximately 1.5 trillion solar masses ($M_\odot$) and a visible matter mass of 250 billion $M_\odot$, in support of a 5:1 dark matter to visible matter ratio. A recent consensus [3] is that the visible baryonic matter of the universe comprises 4.95%, the galactic and peri-galactic dark matter comprises 26.55%, and the dark vacuum energy comprises 68.5% of the critical density (ρ_c), now estimated to be 8.533×10^{-27} kg.m^{-3} (using the 2018 Planck Collaboration H_0 consensus value, 67.4 +/– 0.5 km s^{-1} Mpc^{-1}).

One of the puzzling things about the current cosmological energy density partition of approximately one-third total matter mass-energy and approximately two-thirds dark vacuum energy is the fact that, at present, *these energy densities are of the same order of magnitude! This appears to be an extraordinary coincidence*, because the ΛCDM concordance model *stipulates* that the total matter mass-energy density must have been many orders of magnitude greater than the vacuum energy density in the early universe and will be many orders of magnitude smaller than the vacuum energy density in the future of the universal expansion. How then is it that we happen to be living at just the right time in the history of the universe that these energy densities are nearly equal? Or, *are we wrong in assuming radically different energy density partitions in the remote past and the distant future?* These and other questions arise concerning this unexpected coincidence, which is known among cosmologists as the 'coincidence problem.' As an aside, there is an interesting link between the coincidence problem and the cosmological constant problem, but that is beyond the scope of the current paper. In the history of progress in our understanding of the universe, unexplained coincidences have often been signposts leading to a deeper understanding.

One of the questions one might ask concerning the coincidence problem is whether there could be some systematic error in our quantitation of dark matter that, once corrected, could point to a 50%/50% energy density partition (*i.e.,* 50% total matter mass-energy and 50% dark energy). *This could be achieved, for instance, if approximately one-fifth of the critical energy density currently being classified as a component of the vacuum energy is actually hidden dark matter within the deep intergalactic vacuum.* This has long been one of the predictions of the Flat Space Cosmology (FSC) $R_h = ct$ model [4][5]. In fact, this model stipulates a *perpetual* 50%/50% energy density partition as a falsifiable prediction. This is in stark contrast to the approximately 31.5%/68.5% energy density partition now widely accepted, within tight constraints, and claimed by proponents of the ΛCDM concordance model (see [3]). Over the last 40 years, standard inflationary cosmology has been repeatedly subject to *ad hoc* adjustments from new observations, and has made relatively few falsifiable *predictions* in comparison to the new FSC model [6].

Once-tight constraints, however, can sometimes be broken when new discoveries and/or theories arise. Prior to the discovery of dark energy, for instance, the very idea of a cosmological constant was thought to be highly improbable. A new case in point, as emphasized in this chapter, has been the assumption that the dark matter within and haloed around the galaxies is all, or nearly all, of the discoverable dark matter within the universe.

There is now a new and very plausible theory of dark matter which should loosen the ΛCDM density partition constraints considerably. It is the purpose of this chapter to show how the C.G.S.I.S.A.H. (Cold Ground State InterStellar Atomic Hydrogen) theory of dark matter [7] can be used to identify dark matter within the deep intergalactic vacuum which should be removed from the dark energy ledger column and placed under the total matter mass-energy ledger column.

RESULTS

It has long been known that the intergalactic vacuum is exceedingly sparse with respect to matter. However, it is not entirely empty! Observational 21-cm studies of the cold ground state neutral atomic hydrogen within the intergalactic vacuum, of a similar nature to those made of the interstellar vacuum of the Milky Way, have determined an average intergalactic density of approximately one atom per cubic meter [8]. Thus, the average cold ground state neutral atomic hydrogen density of the intergalactic vacuum is approximately 1.67×10^{-27} kg.m^{-3}. This is one million times less dense than that within the Milky Way galactic disc (1.67×10^{-21} kg.m^{-3}). Nevertheless, one can readily see that the following ratio equation is particularly relevant:

$$[\text{intergalactic H density}]/[\rho_c] = [1.67 \times 10^{-27}\ \text{kg.m}^{-3}]/[8.533 \times 10^{-27}\ \text{kg.m}^{-3}] \quad (1)$$

which equals 0.195.

This is approximately one-fifth of the observed critical density!

DISCUSSION

The ratio equation given in the Results section shows that the currently-observed average density of intergalactic cold ground state neutral atomic hydrogen is 19.5% of the critical density determined from the 2018 Planck Collaboration report [3]. This is a component of the cosmic total matter mass-energy and should not be considered as a component of the vacuum dark energy. By this calculation, one can readily see that 19.5% of the current critical density has been misappropriated by the ΛCDM concordance model as dark energy. In fact, in accordance with the new C.G.S.I.S.A.H. dark matter theory, it should be credited as intergalactic dark matter. Thus, by extending the C.G.S.I.S.A.H. dark matter concept to the intergalactic vacuum, a previously-ignored reservoir of additional dark matter is discovered.

With correct re-apportionment of this additional dark matter to the total matter mass-energy, the cosmic critical energy density partition now becomes approximately 51% total matter mass-energy and 49% dark energy. The current negligible radiation energy contribution to the critical density does not affect the above approximations. Thus, the results of this new C.G.S.I.S.A.H. calculation may be considered to be well within the acceptable margin of observational error, and in support of the FSC prediction of 50%/50%. It should be remembered that there is no cosmological coincidence problem in the FSC model, which stipulates perpetual equality of the cosmological matter mass-energy and non-matter vacuum energy.

SUMMARY AND CONCLUSIONS

The new C.G.S.I.S.A.H. (Cold Ground State InterStellar Atomic Hydrogen) theory of dark matter is applied to the incredibly sparse cold ground state neutral atomic hydrogen within the intergalactic vacuum. Despite an average density of about one atom per cubic meter, this amounts to approximately one-fifth of the cosmic critical density! Thus, according to the new dark matter theory, the intergalactic vacuum contains an additional dark matter contribution to the total matter mass-energy ledger column which must be subtracted from the dark energy ledger column. Within the range of observational error, the resulting 51%/49% matter/dark energy partition of the critical density appears to strongly support the longstanding FSC *prediction* of 50%/50%. There is no cosmological coincidence problem in the FSC model.

Furthermore, this discovery of a universe comprised of 50% matter mass-energy and 50% dark energy suggests that the recent resurrection of the Dirac sea idea [9], as it may pertain to positive and negative *energy* eigenstates (rather than positive and negative *mass* eigenstates) within the FSC model, may be a guide to understanding matter and dark energy as the perpetual yin and yang of universal expansion. This is to be expected in a finite perpetual matter-generating cosmological model which begins from a zero-energy state and follows conservation of energy [10].

REFERENCES

[1] Torres-Flores S., et al. (2011). GHASP: An Hα Kinematic Survey of Spiral and Irregular Galaxies—IX. The NIR, Stellar and Baryonic Tully–Fisher Relations (arXiv:1106.0505)

[2] Gaia-Hubble Collaboration. (2019). http://sci.esa.int/hubble/61198-hubble-and-gaia-accurately-weigh-the-milky-way-heic1905

[3] Aghanim, N., et al. (2018). Planck 2018 Results VI. Cosmological Parameters. http://arXiv:1807.06209v1

[4] Tatum, E.T. (2018). Predicted Dark Matter Quantitation in Flat Space Cosmology. Journal of Modern Physics, 9: 1559–1563. https://doi.org/10.4236/jmp.2018.98096

[5] Tatum, E.T. and Seshavatharam, U.V.S. (2018). Flat Space Cosmology as a Model of Light Speed Cosmic Expansion—Implications for the Vacuum Energy Density. Journal of Modern Physics, 9: 2008–2020. https://doi.org/10.4236/jmp.2018.910126

[6] Tatum, E.T. (2018). Why Flat Space Cosmology Is Superior to Standard Inflationary Cosmology. Journal of Modern Physics, 9: 1867–1882. https://doi.org/10.4236/jmp.2018.910118

[7] Tatum, E.T. (2019). My C.G.S.I.S.A.H. Theory of Dark Matter. Journal of Modern Physics, 10: 881–887. https://doi.org/10.4236/jmp.2019.108058

[8] Cain, F. (2009). What is Intergalactic Space? Universe Today Website https://www.universetoday.com/30280/intergalactic-space

[9] Tatum, E.T. (2019). How the Dirac Sea Idea May Apply to a Spatially-Flat Universe model (A Brief Review). Journal of Modern Physics, 10: 974–979. https://doi.org/10.4236/jmp.2019.108064

[10] Sapar, A. (2019). A Perpetually Mass-Generating Planckian Universe. Proceedings of the Estonian Academy of Sciences, 68(1): 1–12. https://doi.org/10.3176/proc.2019.1.01

CHAPTER 15

How Flat Space Cosmology Models Dark Energy

Abstract: Equations of Flat Space Cosmology (FSC) are utilized to characterize the model's scalar temporal behavior of dark energy. A table relating cosmic age, cosmological redshift, and the temporal FSC Hubble parameter value is created. The resulting graph of the log of the Hubble parameter as a function of cosmological (or galactic) redshift has a particularly interesting sinuous shape. This graph greatly resembles what ΛCDM proponents have been expecting for a scalar temporal behavior of dark energy. And yet, the FSC $R_h = ct$ model expansion, by definition, neither decelerates nor accelerates. It may well be that apparent early cosmic deceleration and late cosmic acceleration both ultimately prove to be illusions produced by a constant-velocity, linearly-expanding, FSC universe. Furthermore, as discussed herein, the FSC model would appear to strongly support Freedman, et al. in the current Hubble tension debate, if approximately 14 Gyrs can be assumed to be the current cosmic age.*

Keywords: Flat Space Cosmology; Dark Energy; Hubble Parameter; Galactic Redshift; $R_h = ct$ model

INTRODUCTION AND BACKGROUND

We are currently in a 'golden age' of astronomy and cosmology. Astrophysical observations in the coming decades are expected to bring much greater resolution concerning the behavior and fundamental nature of dark

*Originally published on October 13, 2020 in Journal of Modern Physics (see Appendix refs).

matter and dark energy. These are two of the remaining great mysteries of the universe.

With respect to the behavior of dark energy, the expansion history of our universe, going back to the earliest galaxies, should come into greater focus. If all goes well with these pending observations, we should be able to fill in many details with respect to the velocities of galactic separation going all the way back to the first few hundred million years of cosmic expansion. We should then have a remarkably accurate 'moving picture' computer simulation of the history of that portion of the universe we can now observe.

When astrophysicists concern themselves with the velocities of galactic separation on scales greater than those of the local clusters held together by gravity and dark matter, they are studying the Hubble parameter and its tight correlation with cosmological redshift. When the Hubble parameter is characterized as a 'snapshot' of the universe at a particular point in cosmic time (at the present time, for instance), it can be referred to as the Hubble *constant*. On a *global* scale, making use of cosmic microwave background (CMB) observations, the 2018 Planck Collaboration has arrived at a *current* Hubble constant H_0 value of 67.36 +/– 0.54 $km.s^{-1}.Mpc^{-1}$ [1].

The ongoing temporal (*i.e.,* 'moving picture') studies of the universe are expected to show that, over the great span of cosmic time, the Hubble parameter is, in fact, *scalar* in some way. The first evidence of this became apparent in 1998, with studies of Type Ia supernovae [2], which revealed the presence of dark energy. Thus, it became apparent that there is an unseen energy, presumably within the cosmic vacuum, which prevents gravitational deceleration of the expanding universe. We now know that universal expansion, at present, is either occurring at constant velocity (as treated by $R_h = ct$ cosmological models) or *very slightly* accelerating (as claimed by ΛCDM concordance model cosmologists). Both types of cosmological models are still viable at the present time [3][4][5][6][7][8]. Observations in the coming decade may well identify which model is superior.

Flat Space Cosmology (FSC) is perhaps the most successful $R_h = ct$ model to date [9]. It *predicts* a current Hubble parameter H_0 value of

66.893 km.s^{-1}.Mpc^{-1}, fitting with the 2018 Planck Collaboration consensus. It also predicts the COBE CMB *dT/T* anisotropy ratio of 0.66×10^{-5}. A book chapter summary of FSC is now freely available online [10]. In contrast to ΛCDM cosmology (which incorporates observations *ad hoc* but makes relatively few falsifiable predictions), the FSC equations provide for very specific predictions, which can falsify the model if proven wrong. Remarkably, to date, the FSC model has not been falsified.

The purpose of this chapter is to show how FSC models the temporal dark energy expansion of the universe. We show in great detail the scalar nature of the FSC Hubble parameter, so that it can be compared to the observations to be made in the coming decades.

METHODS

Previously-published equations of FSC, relating cosmological (or galactic) redshift z, temporal cosmic temperature T_t, temporal cosmic radius R_t, the associated temporal Hubble parameter H_t, the currently-observed Hubble parameter H_0, the currently-observed cosmic temperature T_0, and cosmic age t are brought together in the Results section in order to derive the parameter values given in **Table 1** and **Figure 1** below.

RESULTS

The following two FSC equations are useful for deriving the model relationships between a given cosmological (or galactic) redshift z and the associated temporal Hubble parameter H_t:

$$z \cong \left(\frac{T_t^2}{T_o^2} - 1 \right)^{1/2} \tag{1}$$

and

$$T_t^2 R_t \cong 1.027246639815497 \times 10^{27}\,\mathrm{K}^2 \cdot \mathrm{m} \tag{2}$$

The first equation relates the redshift to the temporal cosmic temperature T_t and the currently-observed cosmic temperature T_0 [11]. The second equation relates the temporal cosmic temperature T_t to the temporal cosmic radius R_t [12].

Recalling the FSC Hubble parameter definition ($H_t = c/R_t$), rearrangement and substitution gives:

$$T_0^2\left(z^2+1\right) \cong H_t\left[\frac{1.027246639815497\times 10^{27}\,\mathrm{K}^2\cdot\mathrm{m}}{c}\right] \quad (3)$$

To convert the H_t term from reciprocal seconds (s^{-1}) to the conventional Hubble parameter units of km.s^{-1}.Mpc^{-1}, the left-hand term is multiplied by $3.08567758 \times 10^{19}$ km.Mpc^{-1}:

$$T_0^2\left(z^2+1\right)\left(3.08567758\times 10^{19}\ \mathrm{km\ Mpc}^{-1}\right) \cong H_t\left[\frac{1.027246639815497\times 10^{27}\,\mathrm{K}^2\cdot\mathrm{m}}{c}\right] \quad (4)$$

Rearrangement of terms gives:

$$\left(z^2+1\right) \cong H_t\left[\frac{1.027246639815497\times 10^{27}\,\mathrm{K}^2\cdot\mathrm{m}}{T_0^2 c\left(3.08567758\times 10^{19}\ \mathrm{km\ Mpc}^{-1}\right)}\right] \quad (5)$$

Using $T_0 = 2.72548$ K, this simplifies to:

$$H_t \cong \frac{\left(z^2+1\right)}{0.014949183831548} \quad (6)$$

The final useful equation relates cosmic time t (in Gyrs after the Planck epoch) to the current Hubble parameter H_0 value of 66.893 km.s^{-1}Mpc^{-1}, the temporal Hubble parameter H_t value, and the current FSC cosmic age of 14.617 Gyrs:

$$H_t \cong H_0\left(\frac{14.617}{t}\right) \tag{7}$$

Equations (5), (6) and (7) can then be used to create **Table 1** and **Figure 1**. The last two z values given in **Table 1** are two of the highest galactic redshifts observed to date.

Notice the *sinuous* appearance of this graph. Its overall shape greatly resembles what cosmologists have been expecting for a scalar temporal behavior of dark energy!

Table 1 Cosmic Age, Redshift z, Hubble Parameter, Log_{10} Hubble Parameter.

Cosmic Age (Gyrs)	Redshift z	H_t (km.s^{-1} Mpc^{-1})	$Log_{10}(H_t)$
14.617	0.00	66.893	1.83
14	0.21	69.84	1.84
13.8	0.24	70.85	1.85
13	0.35	75.21	1.88
12	0.47	81.48	1.91
11	0.57	88.89	1.95
10	0.68	97.78	1.99
9	0.79	108.64	2.04
8	0.91	122.22	2.09
7	1.04	139.68	2.15
6	1.20	162.96	2.21
5	1.39	195.55	2.29
4	1.63	244.44	2.39
3	1.97	325.92	2.51
2	2.51	488.89	2.69
1	3.69	977.77	2.99
0.5	5.31	1955.55	3.29
0.25	7.58	3911.1	3.59
0.174	9.11	5618.51	3.75
0.1179	11.09	8293.97	3.92

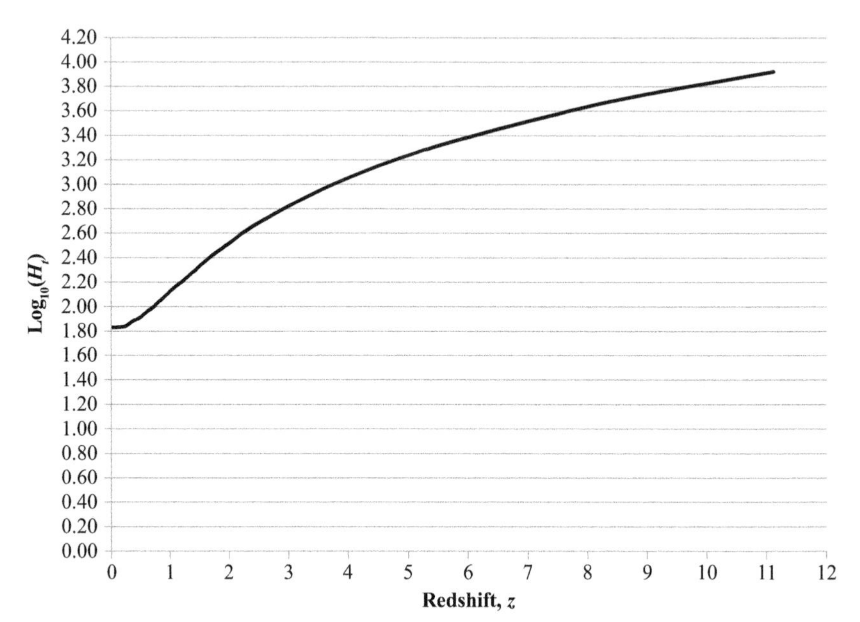

Figure 1 $\mathrm{Log}_{10}(H_t)$ as a Function of Cosmological (or Galactic) Redshift z.

DISCUSSION

Proponents of the ΛCDM concordance model of cosmology, and $R_h = ct$ model cosmologists, are currently in a pitched battle to establish which model is more accurate with respect to observations and predictions. As documented in recent publications [13][14], FSC is a realistic linear light-speed cosmic expansion model which can also be considered a modified Milne 'empty universe' model. Following a sign convention which treats gravitationally-attracting matter energy density as positive and 'repulsive gravity' vacuum energy density as negative, the FSC *net* global energy density is perpetually zero. Thus, the FSC cosmic model follows the 'empty' line demarcating deceleration and acceleration in Figure 2 of the following link: https://dx.doi.org/10.4236/jmp.2020.1110091

The 'empty' universe line in this open-source figure published by the Supernova Cosmology Project (SCP) [15] falls exactly where the FSC and other R_h=ct models fall. One can also readily see how it is that R_h=ct universe models appear to show an excellent statistical fit with SCP observations to date. *Clearly, the observational error bars allow for BOTH models [i.e., the blue line of ΛCDM accelerating expansion, as well as the 'empty'* *pink line corresponding to constant velocity expansion of the FSC $R_h = ct$ model]*. Notice also that the SCP figure correlates a redshift z value of 1.0 with a cosmic scale of 0.5 times the current scale. This is true for FSC as well as ΛCDM, although the two models differ slightly with respect to the current cosmic age.

In ΛCDM cosmology, the post-inflationary cosmological vacuum energy density is assumed to be a constant. This is not an absolute requirement of general relativity, so long as the vacuum energy density is scalar according to $\Lambda = 3H_t^2/c^2$. In the FSC quintessence model, this scalar relationship holds true and is equivalent to $\Lambda = 3/R_t^2$ [16]. In FSC, the vacuum energy density declines in the forward time direction approximately 121 logs of 10 from the Planck scale epoch to the present. Thus, in contrast to ΛCDM cosmology, there is no 'cosmological constant problem' in the FSC model.

As speculated in Chapter one, ongoing cosmological positive-energy matter creation might be paired with an ongoing increase in cosmological negative vacuum energy, as a requirement for conservation of energy in such a finite isolated expanding system. It should be remembered that the details of matter creation in *all* cosmological models are a mystery. In FSC, matter creation is treated as an ongoing process, whereas ΛCDM cosmologists generally assume that all matter was created nearly instantaneously. However, as a result, a major difference between the two models is that *only ΛCDM cosmology has a cosmological constant problem, based upon its embedded constant post-inflationary vacuum energy density assumption.*

As a consequence of the dark energy observations, in addition to their cosmological constant and instantaneous matter creation assumptions, ΛCDM cosmologists must now also assume certain features of the universal expansion. These features had not been required when it was once thought (*i.e.*, before 1998) that the cosmological vacuum energy density might actually be perpetually zero. They now require that universal expansion decelerated during the first half of the cosmic time span since the Big Bang, and then, almost imperceptibly, began to accelerate approximately 6 billion years ago. This becomes absolutely necessary if one requires a post-inflationary cosmological *constant* at the currently observed value of about 10^{-9} Joules per cubic meter. Nevertheless, this deceleration-followed-by-acceleration scenario of universal expansion is clearly debatable, especially when one considers the observational statistical error bars in the SCP figure.

When one compares the relative luminosity and angular diameter distances between the two competing models, in the form of a ratio, it has recently been shown that the ΛCDM model contention of late cosmic acceleration could be an illusion produced by a $R_h = ct$ universe [17] (see Chapter four).

Further support that cosmic acceleration could be an illusion is clearly evident in **Figure 1** of the current report. It is readily apparent that the FSC graph of the log of the Hubble parameter as a function of redshift z is *sinuous* in appearance. We see the following: an upward flexion curve out to a z value of about 1.0 (corresponding to the last 7.3 billion years of the FSC cosmic expansion); a roughly straight line segment for $1.0 < z < 1.7$ (corresponding to 3.76 to 7.3 billion years of cosmic age); and an opposite flexion curve for z values greater than about 1.7 (corresponding to the first 3.76 billion years of the FSC cosmic expansion). The overall shape of the graph greatly resembles what ΛCDM proponents have been expecting for a scalar temporal behavior of dark energy. And yet, the FSC $R_h = ct$ model expansion, by definition, neither decelerates nor accelerates!

The upward curving portion of our **Figure 1** graph out to a z value of about 1.5, is already largely filled in by the accumulated Type Ia supernovae data [18]. Not yet known are the *exact* Hubble parameter values at the cosmic times when these supernovae exploded. Fortunately, the coming decade of observational studies should give us a better idea of the precise scalar nature of the Hubble parameter.

Regardless, given the overall shape of our **Figure 1** graph, it may well be that *apparent* early cosmic deceleration and late cosmic acceleration *both* ultimately prove to be illusions produced by a constant-velocity, linearly-expanding, FSC universe.

Given the ongoing tension between different research teams considering what *current* near and deep space observations might be telling us about the H_0 value as a snapshot in time, it is worth noting the following:

The 2018 Planck Collaboration analysis of the CMB looked at 99.998 percent of the current radius of the universe. Their consensus H_0 estimate of 67.36 $km.s^{-1}.Mpc^{-1}$ appears, in FSC, to fit with a 14.6 Gyr old universe.

According to **Table 1**, the Freedman, et al H_0 observation of 69.6 km.s^{-1}.Mpc^{-1} [19] appears to be fitted nicely to a 14 Gyr estimated cosmic age. Whereas, the SHoES project H_0 observations of 74–77 km.s^{-1}.Mpc^{-1} [20] appear to be ideally fitted to a 13 Gyr (or less) cosmic age. One need only consider the current 14.27 +/– 0.38 Gyr best age estimate of the HD 140283 'Methuseleh star' [21] to judge which current H_0 estimate is the most likely outlier.

SUMMARY AND CONCLUSIONS

Equations of FSC have been utilized to characterize the model's temporal behavior of dark energy. A table relating cosmic age, cosmological redshift, and the temporal FSC Hubble parameter value has been created. The resulting graph of the log of the Hubble parameter as a function of cosmological (or galactic) redshift has a particularly interesting sinuous shape: an upward flexion curve out to a z value of about 1.0 (corresponding to the last 7.3 billion years of the FSC cosmic expansion); a roughly straight line segment for $1.0 < z < 1.7$ (corresponding to 3.76 to 7.3 billion years of cosmic age); and an opposite flexion curve for z values greater than about 1.7 (corresponding to the first 3.76 billion years of the FSC cosmic expansion). The overall shape of the graph greatly resembles what ΛCDM proponents have been expecting for a scalar temporal behavior of dark energy. And yet, the FSC $R_h = ct$ model expansion, by definition, neither decelerates nor accelerates. It may well be that apparent early cosmic deceleration and late cosmic acceleration both ultimately prove to be illusions produced by a constant-velocity, linearly-expanding, FSC universe.

REFERENCES

[1] Aghanim, N., et al. (2018). Planck 2018 Results VI. Cosmological Parameters. arXiv:1807.06209v1

[2] Perlmutter, S., et al. (1999). Astrophysical Journal, 517: 565–586. https://doi.org/10.1086/307221.

[3] Melia, F. (2012). Fitting the Union 2.1 SN Sample with the Rh = ct Universe. Astronomical Journal, 144: arXiv:1206.6289 [astro-ph.CO]

[4] Nielsen, J.T., et al. (2015). Marginal Evidence for Cosmic Acceleration from Type Ia Supernovae. Scientific Reports, 6: Article number 35596. doi:10.1038/srep35596. arXiv:1506.01354v1.

[5] Jun-Jie Wei, et al. (2015). A Comparative Analysis of the Supernova Legacy Survey Sample with ΛCDM and the R_h = ct Universe. *Astronomical Journal*, 149: 102.

[6] Tutusaus, I., et al. (2017). Is Cosmic Acceleration Proven by Local Cosmological Probes? Astronomy & Astrophysics, 602_A73. arXiv:1706.05036v1 [astro-ph.CO].

[7] Dam, L.H., et al. (2017). Apparent Cosmic Acceleration from Type Ia Supernovae. Mon. Not. Roy. Astron. Soc. arXiv:1706.07236v2 [astro-ph.CO]

[8] Melia, F. (2019). Tantalizing New Physics from the Cosmic Purview. doi:10.1142/S0217732319300040 https://arxiv.org/abs/1904.11365

[9] Tatum, E.T. (2018). Why Flat Space Cosmology Is Superior to Standard Inflationary Cosmology, Journal of Modern Physics, 9: 1867–1882. https://doi.org/10.4236/jmp.2018.910118

[10] Tatum, E.T. (2020). A Heuristic Model of the Evolving Universe Inspired by Hawking and Penrose. In Eugene T. Tatum (Ed.). *New Ideas Concerning Black Holes and the Universe*. (pp. 5–21). London: IntechOpen. http://dx.doi.org/10.5772/intechopen.87019

[11] Tatum, E.T. and Seshavatharam, U.V.S. (2015). Flat Space Cosmology as a Mathematical Model of Quantum Gravity or Quantum Cosmology. International Journal of Astronomy and Astrophysics, 5: 133–140. http://dx.doi.org/10.4236/ijaa.2015.53017

[12] Tatum, E.T. and Seshavatharam, U.V.S. (2018). Temperature Scaling in Flat Space Cosmology in Comparison to Standard Cosmology. Journal of Modern Physics, 9, 1404–1414. https://doi.org/10.4236/jmp.2018.97085

[13] Tatum, E.T. (2019). How the Dirac Sea Idea May Apply to a Spatially-Flat Universe Model (A Brief Review). Journal of Modern Physics, 10: 974–979. https://doi.org/10.4236/jmp.2019.108064

[14] Tatum, E.T. and Seshavatharam, U.V.S. (2018). Flat Space Cosmology as a Model of Light Speed Cosmic Expansion - Implications for the Vacuum Energy Density. Journal of Modern Physics, 9: 2008–2020. https://doi.org/10.4236/jmp.2018.910126

[15] Perlmutter, S. (2016). Supernova Cosmology Project. Lawrence Berkeley National Laboratory, Berkeley. http://www.supernova.lbl.gov

[16] Tatum, E.T. and Seshavatharam, U.V.S. (2018). Clues to the Fundamental Nature of Gravity, Dark Energy and Dark Matter. Journal of Modern Physics, 9: 1469–1483. https://doi.org/10.4236/jmp.2018.98091

[17] Tatum, E.T. and Seshavatharam, U.V.S. (2018) How a Realistic Linear Rh = ct Model of Cosmology Could Present the Illusion of Late Cosmic Acceleration. Journal of Modern Physics, 9: 1397–1403. https://doi.org/10.4236/jmp.2018.97084

[18] Betoule, M., et al. (2014). Improved Cosmological Constraints from a Joint Analysis of the SDSS-II and SNLS Supernova Samples. A & A, 568: A22. https://doi.org/10.1051/0004-6361/201423413

[19] Freedman, W.L., et al. (2020). Calibration of the tip of the Red Giant Branch. Astrophysical Journal, 891(1): 57–75. doi:10.3847/1538-4357/ab7339

[20] Reiss, A.G., et al. (2016). A 2.4% Determination of the Local Value of the Hubble Constant. arXiv:1604.01424v3. doi:10.3847/0004-637X/826/1/56

[21] Vandenberg, D.A., et al. (2014). Three Ancient Halo Subgiants: Precise Parallaxes, Compositions, Ages, and Implications for Globular Clusters. Astrophysical Journal, 792(2): art. No. 110. https://doi.org/10.1088/0004-637X/792/2/110

CHAPTER 16

Dark Matter as Cold Atomic Hydrogen in its Lower Ground State

Abstract: This chapter expands on a novel theory of dark matter made plausible by several astronomical observations reported in 2018 and 2019. One of us (E.T.T.) introduced his theory to colleagues invited to the Dark Matter Workshop at the World Science Festival in May of 2019, and its first publication was in a peer-reviewed physics journal in July of 2019.*

Keywords: Dark Matter; Atomic Hydrogen; Interstellar Medium; Cosmic Dawn; Wouthuysen-Field Effect; Cosmology Theory; Milky Way Galaxy

INTRODUCTION AND BACKGROUND

The theory [1], simply stated, is that what we currently refer to as 'cold dark matter' is, in actuality, slow-moving interstellar and intergalactic neutral atomic hydrogen in its lower 1s ground state. Its exceedingly low density within the vacuum of space can be quantified by measuring the intensity of its signature spectral hyperfine 21-cm *absorption* line in lines-of-site to stellar objects at known distances. At an average HI density of approximately one atom per cubic centimeter (1.67×10^{-21} kg.m^{-3}) within the vast, cold and remote interstellar vacuum of the Milky Way, it is very nearly collision-less and thus mostly unperturbed. And, given its current nearly perpetual lower ground state condition, it *cannot* emit light.

*Originally published on March 20, 2020 by IntechOpen (see Appendix refs).

Whenever and wherever hydrogen is mostly above this ground state, and significantly more concentrated, it is readily visible and we call it something else (a cold, warm or hot gas cloud, for instance).

Following a brief review of the historical evidence for the existence of dark matter, its key observations reported in 2018 and 2019 will be summarized and its current constraints elaborated. The author's calculations, in the context of these observations, will then be presented in the Results section, and a Discussion section with a table based upon these findings will follow.

HISTORICAL EVIDENCE FOR DARK MATTER

It is generally agreed that astronomer Fritz Zwicky, in 1933, was the first scientist to apply the virial theorem to infer the existence of dark matter. He referred to it as 'dunkle materie' [2][3]. Unfortunately, Zwicky's dark matter proposal was largely ignored at the time.

Beginning in 1970, this problem of 'missing matter' was further elucidated and essentially proven by the detailed studies of galactic rotation by Vera Rubin and William Ford [4][5], although it took considerable time for them to receive due recognition for this achievement.

With gradual acceptance of the observational implications, what has followed in the ensuing decades has been a stepwise progression of tightening constraints on the nature and quantity of dark matter. As a consequence, much like a horse race with changing leads, various creative and exotic theories of the nature of dark matter (WIMPs, MACHOs, axions, sterile neutrinos, supersymmetry partners, SIMPs, GIMPs, etc.) have fallen in and out of favor [6]. Given the difficulty of its detection, there have even been attempts to discard the idea of dark matter altogether in favor of modifying Newtonian celestial mechanics (Modified Newtonian Dynamics, or MOND) [7][8].

A review of these various theories, and a discussion of their current plausibility, is beyond the scope of this chapter. Whole books have been written about them. Suffice it to say, in view of the many continuing exotic dark matter detector failures, there is room for a new theory such as the one presented herein. The following section will summarize key constraints on dark matter as of 2020.

CURRENT OBSERVATIONAL CONSTRAINTS ON DARK MATTER

Upon establishing the likelihood of an abundance of cosmic matter which, *in its current state*, does not emit light, astronomers and astrophysicists have attempted to quantify it with respect to the visible matter (*i.e.,* stars, gas clouds and cosmic dust). The 2018 Planck Collaboration report [9] indicates a cosmic dark matter-to-visible matter ratio of approximately 5.4-to-1. This is in close agreement with a ratio of approximately 5-to-1 established by a 2019 *Gaia*-Hubble survey report [10] on the Milky Way (MW) galaxy. The *Gaia* report indicates a total virial MW mass of approximately 1.5 trillion solar masses which includes a visible matter mass of approximately 250 billion solar masses. Based upon these and other studies, dark matter is currently believed to comprise about 85% of all cosmic matter. Thus, although it appears, by gravitational lensing, to be predominantly within and haloed around the visible galaxies, dark matter is most likely ubiquitous and therefore a key structural (*i.e.,* 'scaffold') component of the universe. In this context, it is worth noting that the Planck Collaboration study of the cosmic microwave background (CMB) anisotropy documents the presence and gravitational influence of dark matter within the hot and dense early universe during the recombination/decoupling epoch. So what we now tend to think of as 'cold dark matter' (CDM) was once hot, and very possibly light-emitting, in its past excited state.

Although relatively few in number, MW halo stars at various known distances beyond the galactic disk can provide for line-of-site spectral analysis and a rough MW halo vacuum density determination of interstellar neutral atomic hydrogen in its lower ground state. Specifically, the intensity of the hyperfine 21-cm absorption line gives us some idea of the number of these particular atoms per unit volume of the column of intervening interstellar space. Best estimates of this sort, made over a number of decades, have indicated an average density within the MW interstellar vacuum of roughly one of these atoms per cubic centimeter [11][12][13].

Making use of some initial *Gaia* survey data released in 2018, Posti and Helmi reported results [14] which allow one to deduce a ratio of dark matter-to-visible matter within a 20 kpc (*i.e.,* 65 thousand light-years) radius halo sphere of the MW (see schematic **Figure 1**). This halo sphere

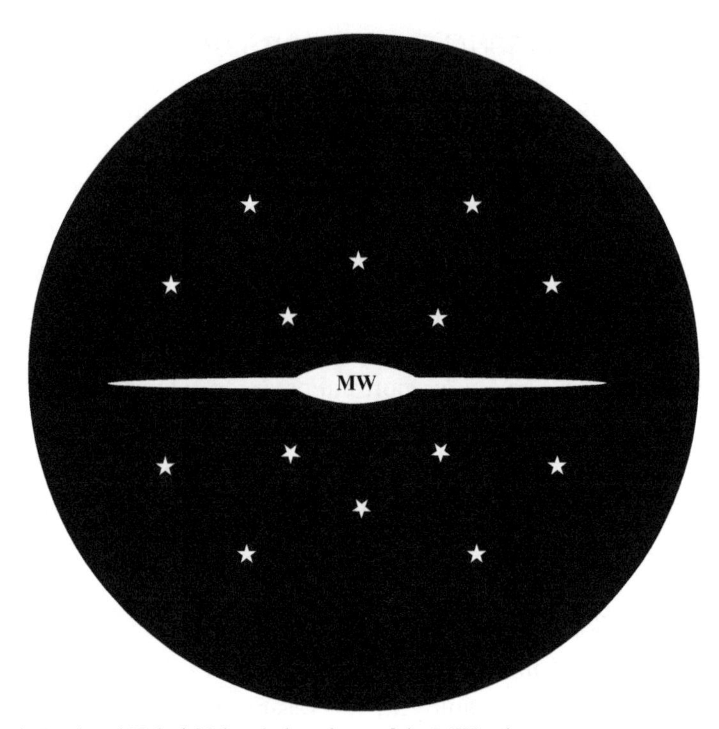

Figure 1 Posti and Helmi 20 kpc halo sphere of the MW galaxy.

is represented in black in the figure and is roughly to scale with respect to the 50 thousand light-year radius MW disk (in white). The disk averages approximately one thousand light-years in thickness. The relatively few halo stars well beyond the disk are also schematically represented in the figure. As mentioned, these are useful for density measurements of cold hydrogen in the lower ground state within the halo vacuum.

The total virial mass of their sphere was reported by Posti and Helmi to be 1.91×10^{11} $M_{\odot}$ (solar masses), of which the mass of dark matter was reported to be 1.37×10^{11} $M_{\odot}$. This would imply that the MW 20 kpc sphere ratio of dark matter-to-visible matter is about 2.54-to-1. Therefore, if we normalize the MW visible mass to the 250 billion $M_{\odot}$ value given in the 2019 *Gaia* survey report, this Posti and Helmi ratio would imply a corresponding dark matter mass of approximately 635 billion $M_{\odot}$ within the same 20 kpc radius halo sphere. These numbers will be compared in the subsequent Results section.

Aside from the inability of dark matter to emit light, observations have confirmed that it is nearly collision-less. It appears to be composed of particles with a low scattering cross-section. This can be deduced from Tucker's early observations of the bullet cluster [15] and subsequent observations of other colliding galaxies.

Dark matter, at present, is also believed to be cold (*i.e.,* slow-moving). A predicted Maxwell-Boltzmann particle velocity distribution ranging from roughly 0–600 km/sec, and peaking at roughly 220–230 km/sec, is the theoretical basis for optimizing a variety of cold dark matter particle detectors [16]. Unfortunately, none of these experiments to date has produced a positive result of an exotic (*i.e.,* non-baryonic) dark matter particle. Intriguingly, however, the 2018 EDGES study [17] of the hyperfine 21-cm spectral line of neutral atomic hydrogen corresponding to cosmological redshifts of $15 < z < 20$ (cosmic dawn) has reported a strong signal consistent with a hydrogen gas temperature in the low single digits of the Kelvin temperature scale. This is considerably lower than the cosmic dawn CMB radiation temperature and produces strong constraints on the nature of dark matter. This *CMB decoupling phase* during cosmic dawn indicates that whatever we are currently referring to as dark matter has been particularly cold since at least the time of early cosmic dawn, has a particle mass of no more than about 2-3 GeV, and has a scattering cross-section σ_1 value of at least 1.5×10^{-21} cm^2. If the EDGES observations of cosmic dawn are, in fact, the result of dark matter cooling of warmer (*i.e.,* CMB-equilibrated) hydrogen atoms, the proposed WIMPs and all but one baryon (namely, *colder* atomic hydrogen in its lower ground state) are effectively ruled out as dark matter candidates.

Figure 3 on page 9 of Barkana's review [18] related to the EDGES study findings summarizes the new cosmic dawn dark matter constraints with a log graph of the implied b-DM cross-section σ_1 and the minimum possible 21-cm brightness temperature (T_{21}) on the two vertical axes and the corresponding implied dark matter particle mass M_X on the horizontal axis. All constraint values indicated in the graph correspond to the strong signal measured at $z = 17$, which corresponds to a redshifted 21-cm hyperfine hydrogen absorption line detectable at a frequency of 78.9 MHz. To fully comprehend the significance of these dark matter constraints, the reader should obtain this reference and pay particular attention to the dark matter particle mass corresponding to a cross-section σ_1 value of 10^{-20} cm^2 and a 21-cm brightness temperature $\log_{10}$ value (in mK) of 2.32. Please note that *these values correspond to a cold dark matter particle fitting with neutral atomic hydrogen, which has a similar low velocity scattering cross-section and a mass-energy of 0.938 GeV. Furthermore, it should be remembered that the 21-cm absorption line is the signature of atomic hydrogen in its lower ground state. These new cosmic dawn constraints on dark matter will be a major focus in the following Discussion section, particularly with respect to the Wouthuysen-Field effect.*

Without specifically naming any particular non-excluded baryons, physicist Stacy McGaugh published a brief note [19] at the time of the EDGES publication (March, 2018) which strongly supports the idea that the cosmic dawn observations are to be, in his words, '*expected for a purely baryonic universe.*' He begins the note with the observation that the intensity of the redshifted hyperfine 21-cm spectral line at $z = 17$ is anomalously strong for ΛCDM, which proposes non-baryonic dark matter. He also points out that current knowledge in atomic physics would indicate that a maximum intensity T_{21} signal should occur when the neutral hydrogen fraction X_{HI} equals 1 and the spin temperature T_S equals the kinetic temperature T_K of the primordial gas. McGaugh's cogent arguments and interpretation of the EDGES cosmic dawn data are strongly supportive of the theory presented herein.

An additional constraint on dark matter has to do with the 'cusp-core problem,' specifically why some galaxies have a distinctly cuspy distribution of dark matter and others do not. A 2019 report on dark matter distribution within dwarf galaxies, by Read, et al. [20], offers a clue. It shows that galaxies which stopped forming stars over 6 billion years ago tend to be cuspier than those with more extended star formation. This is equivalent to saying that the extended star formation dwarf galaxies have shallower dark matter cores. Thus, their findings agree well with models where dark matter is presumably heated up by bursty star formation. This means that any plausible theory of dark matter must explain why extended and bursty star formation is correlated with a so-called 'cored' dark matter distribution.

One obvious possible interpretation of the Read observations is simply that bursts of highly energetic particles and photons, produced by a concentration of new stars in and around active galactic centers, would tend to heat up and eject cold dark matter from their vicinity. If this is the correct interpretation, then a self-interacting dark matter (SIDM) model becomes unnecessary to explain the 'cusp-core problem.' In fact, all sorts of bizarre non-baryonic properties of dark matter then become unnecessary.

RESULTS (CALCULATION)

Given the new dark matter theory as briefly summarized in the Introduction section, a simple calculation can be made on the Posti and Helmi 20 kpc MW halo sphere, as a test of this theory. If we start with the current best estimate of an average of only one atom of atomic hydrogen in the lower ground state per cubic centimeter of the Posti and Helmi 20 kpc halo sphere, that assumes a vacuum hydrogen density of 1.67×10^{-21} kg.m^{-3}. If we then multiply that number by the volume of the 20 kpc sphere (9.85×10^{62} m^3), the total mass of atomic hydrogen in the bottom ground state is 1.645×10^{42} kg. That is the equivalent of 827 billion $M_{\odot}$. This is 3.3 times the 2019 *Gaia* survey MW galaxy visible mass! Even allowing for only 0.75 atom of atomic hydrogen in the bottom ground state per cubic centimeter of the 20 kpc halo sphere, the Posti and Helmi dark matter-to-visible matter ratio of 2.54 can be met.

DISCUSSION: INTERSTITIAL HYDROGEN, COSMIC DAWN AND THE WOUTHUYSEN-FIELD EFFECT

Observations of the CMB anisotropy map suggest the following cosmic evolution scenario since the CMB emission epoch:

Denser regions of the primordial hydrogen distribution, already subject to the positive feedback of gravity, further aggregated into the hot stars, warm gas clouds, galaxies, quasars and filaments. In contrast, due to adiabatic cosmic expansion, the primordial hydrogen within the low gravity interstices of the CMB map progressively became exceedingly sparse and cold (*i.e.*, CMB-equilibrated). These interstices we know today as the vast interstellar and intergalactic space, including the voids.

The expanding and cooling universe, after CMB emission, was completely dark before the first dense clusters of primordial hydrogen underwent nuclear fusion. This period, known as the cosmic 'dark age,' merged into the 'cosmic dawn' reionization epoch at around 100 million years after the big bang. The 'cosmic dawn' epoch is named as such because this is when the first stars are thought to have formed.

As documented by the EDGES study, a strange phenomenon occurred during the period of cosmic dawn. For about 150 million years, corresponding roughly to the cosmological redshift range of $15 < z < 20$, the temperature T_G of the vast interstitial primordial hydrogen gas was *decoupled* from the CMB radiation temperature T_R. At the peak of this phenomenon, at roughly $z = 17$, this primordial hydrogen appears to have been in the low single digits of the Kelvin temperature scale. Thereafter, the hydrogen gas gradually warmed back up to the CMB temperature at roughly $z = 15$. **Figure 2** illustrates this phenomenon. On this graph $z = 20$ corresponds to about 100 million years after the big bang, $z = 17$ corresponds to about 180 million years after the big bang, and $z = 15$ corresponds to about 250 million years after the big bang.

This phenomenon of 'cosmic dawn CMB decoupling' is most commonly attributed to a baryon-dark matter (b-DM) scattering interaction, whereby dark matter is presumed to have chilled faster than primordial

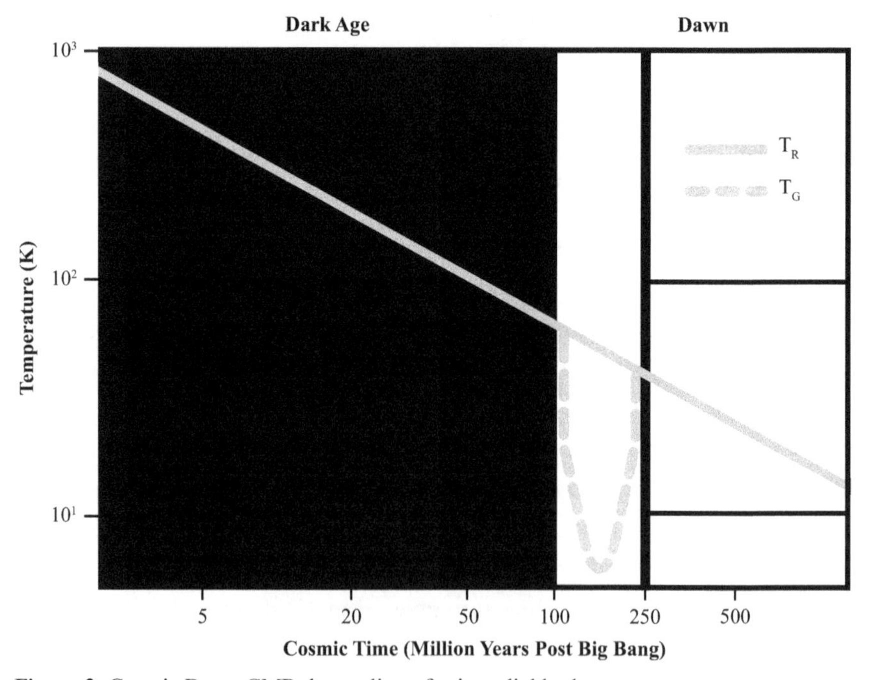

Figure 2 Cosmic Dawn CMB decoupling of primordial hydrogen.

hydrogen during the cosmic dark age, to the point where it could then interact with and chill the CMB-equilibrated interstitial hydrogen and decouple it from the CMB radiation temperature.

The problem with this particular explanation of the EDGES study observations is to explain why the beginning of the CMB decoupling phenomenon *coincided with the first stars* at the crack of cosmic dawn. How is it that dark matter had cooled sufficiently to enable b-DM scattering and CMB decoupling of primordial hydrogen just when the first stars were forming? Could there be a simpler explanation for cosmic dawn CMB decoupling *without requiring a non-baryonic intermediary?*

Fully in keeping with McGaugh's bold assertion of a purely baryonic mechanism, this cosmic dawn coincidence may have been *entirely* due to the Wouthuysen-Field (WF) effect on CMB-equilibrated primordial atomic hydrogen. If unfamiliar with this radiation effect on atomic hydrogen, the reader is encouraged to read an excellent and brief summary of the WF effect on the Wikipedia page entitled 'Wouthuysen-Field Coupling' [21]. A more extensive and highly technical summary is also found on the Astro-Baki website [22]. Briefly, the Lyman-alpha ultraviolet (UV) radiation of the first stars was of sufficient energy to have caused a redistribution of the balance of the two hydrogen electron hyperfine 21-cm ground states such that the primordial hydrogen gas could effectively bypass its 'forbidden transition' (from parallel to antiparallel electron spin) and *easily reach the lower ground state.* The net effect of this process would have been to decouple primordial hydrogen from the CMB radiation temperature, producing the strong 21-cm absorption line signal observed. Thus, it appears that *an exotic, non-baryonic, form of dark matter was completely unnecessary for cosmic dawn CMB decoupling. The mysterious dark matter at cosmic dawn could simply have been the first of the interstitial hydrogen to be chilled and decoupled by the Lyman-alpha radiation. The process then, over millions of years, would have extended to the rest of the CMB-equilibrated hydrogen, peaking at a cosmic redshift of $z = 17$.*

The key dark matter features, including observational constraints achieved over the last few years, and the correlating features of interstitial atomic hydrogen in the lower HI ground state, can now be brought together in a table (**Table 1**) for comparison:

Table 1 Dark Matter Features vs Interstitial HI Cold Hydrogen.

Dark Matter Features	Interstitial HI Cold Hydrogen	Ref
Cold (0–600 km/sec)	Cold (0–600 km/sec)	[16]
Dark (no emissions)	Lower Ground State (cannot emit)	[2][3][4][5]
Cross-Section $\sigma_1 > 1.5 \times 10^{-21}$ cm^2	$\sigma_1 > 1.5 \times 10^{-21}$ cm^2 (at low velocity)	[15][17][18]
Baryon (strongest 21-cm signal)	Baryon for $X_{HI} = 1$ and $T_S = T_K$	[19]
Mass-Energy less than 3 GeV	Mass-Energy = 0.938 GeV	[17][18]
Mass 20 kpc Halo = 635 Billion $M_\odot$	Mass 20 kpc Halo = 827 Billion $M_\odot$	[14][10]
Central DM Heating ('coring')	Ejected and Loses Ground State	[20]
CMB Decoupling at Cosmic Dawn	Wouthuysen-Field Effect	[21][22]
Structural Scaffold	Most Abundant Atom	[9]
Existence at CMB Emission	Most Abundant Atom	[9]

These correlations are striking and *strongly suggest that interstitial cold atomic hydrogen in its lower ground state is what we have been calling dark matter over the last few decades.*

It has long been assumed that the average atomic density of the 'nearly empty' vacuum of interstellar space beyond the visible stars, gas clouds and cosmic dust can be ignored in galactic mass calculations. While this might be true for the confines of the galactic disk and bulge, where visible matter is particularly concentrated, it is definitely not true for the galactic halo in close proximity to the disc. The sheer vastness of space belies the assumption mentioned above. It appears that this mistaken assumption has been a key foundational error behind the longstanding mystery of dark matter. The simple calculation in the Results section supports this conclusion. Even a single stray baryonic atom per cubic centimeter of interstellar space within the 20 kpc MW halo of Posti and Helmi can dwarf the combined mass of all visible stars, clouds of gas and cosmic dust!

The fact that the particular atom in question appears now not to be in the least bit exotic but, instead, the most common structural element in the universe, is indeed ironic. In a sense, because of the many distractions and obscurations provided by the highly visible warm and hot hydrogen atoms, cold interstitial hydrogen, because of its remote location, extremely low density, low velocity, and prolonged lower ground state, has been *essentially hiding in plain sight.* Observations of the 21-cm hyperfine *absorption* line (its *signature*) have been noted for decades but only recently connected to phenomena attributed to dark matter.

Any useful physical theory should be falsifiable and predictive. The falsifiability of this particular theory is obvious. This theory would be falsified if a particle M_X of 0.938 GeV becomes excluded from dark matter constraints, or current best estimates of the average MW halo vacuum density of cold atomic hydrogen are subsequently proven to be *severely* overestimated. However, a minor correction to approximately 0.5–0.75 atom per cubic centimeter is entirely consistent with this theory. As for observations to further strengthen this theory, the following predictions are made:

1. There will be tightening dark matter constraints around a particle M_X value of 0.938 GeV (*i.e.,* the mass-energy of neutral atomic hydrogen).
2. Computer simulations of galaxy formation and evolution which incorporate this theory will show excellent correlations with observations, including the coring effect of heating and ejecting cold interstellar hydrogen from active galactic centers with bursty star formation.
3. No exotic non-baryonic particles fitting the observed qualitative and quantitative constraints will ever be discovered.

SUMMARY AND CONCLUSION

To summarize, this chapter has introduced the reader to a plausible new theory of dark matter which appears to match current observational constraints. The theory, simply stated, is that what we currently refer to as 'cold dark matter' is, in actuality, slow-moving interstellar and intergalactic neutral atomic hydrogen in its lower 1s ground state. So long as it stays in this lower ground state, it cannot emit light. Furthermore, it is currently so sparse as to be nearly collision-less. Whenever and wherever hydrogen is mostly above this ground state, and significantly more concentrated, it is readily visible and we call it something else (a cold, warm or hot gas cloud, for instance).

Dark matter observations corresponding to the cosmic dawn epoch, which were reported in 2018 and 2019, have provided the necessary constraints on dark matter to favor this theory above all others at the present time. In particular, the Bowman (*i.e.,* EDGES) and Barkana references point to a cold dark matter particle with features quite consistent with cold atomic hydrogen. Furthermore, a convincing case has been made by McGaugh that the intense 21-cm hydrogen absorption signal at cosmic dawn is the *signature* of a baryonic universe. The obvious mechanism for such signal strength, and its coincidence with cosmic dawn, is the Wouthuysen-Field effect. From the forgoing discussion, it becomes apparent that exotic (*i.e.,* non-baryonic) matter is not necessary to explain dark matter observations to date.

We conclude by asking the following question:

'If interstitial cold atomic hydrogen in its lower ground state is qualitatively and quantitatively sufficient to explain dark matter observations to date, do we really need to spend more of our time and money continuing to look for anything else?'

REFERENCES

[1] Tatum, E.T. (2019). My C.G.S.I.S.A.H. Theory of Dark Matter. Journal of Modern Physics. 10: 881–887. https://doi.org/10.4236/jmp.2019.108058
[2] Zwicky, F. (1933). Die Rotverschiebung von extragalaktischen Nebeln. Helvetica Physica Acta. 6: 110–127, Bibcode:1933AcHPh...6..110Z
[3] Zwicky, F. (1937). On the Masses of Nebulae and of Clusters of Nebulae. Astrophysical Journal. 86: 217, Bibcode:1937ApJ....86..217Z, doi: 10.1086/143864
[4] Rubin, V. and Ford, W.K. (1970). Rotation of the Andromeda Nebula from a Spectroscopic Survey of Emission Regions. The Astrophysical Journal. 159, 379 (doi: 10.1086/150317)
[5] Rubin, V., et al. (1985). Rotation Velocities of 16 SA Galaxies and a Comparison of Sa, Sb, and SC Rotation Properties. The Astrophysical Journal. 289, 81 (doi: 10.1086/162866)
[6] Hadhazy, A. (2019). The Dark Matter Derby. Discover. 40(9): 40–47.

[7] Milgrom, M. (1983). A Modification of the Newtonian Dynamics as a Possible Alternative to the Hidden Mass Hypothesis. Astrophysical Journal. 270: 365–70. Bibcode:1983ApJ...270..365M. doi: 10.1086/161130

[8] Kroupa, P., Pawlowski, M. and Milgrom, M. (2012). The Failures of the Standard Model of Cosmology Require a New Paradigm. International Journal of Modern Physics D. 21(14):1230003. arXiv:1301.3907. Bibcode:2012IJMPD..2130003K. doi: 10.1142/S0218271812300030

[9] Aghanim, N., et al. (2018). Planck 2018 Results VI. Cosmological Parameters. http://arXiv:1807.06209v1

[10] Watkins, L.L., et al. (2019). Evidence for an intermediate-mass Milky Way from *Gaia* DR2 Halo Globular Cluster Motions. The Astrophysical Journal. 873: 118–130. doi: 10.3847/1538-4357/ab089f

[11] Mammana, D.L. *Interstellar Space*. New York: Popular Science, 2002 (p. 220).

[12] Chaisson, E., S. McMillan. *Astronomy Today*. New York: Prentice Hall, 1993 (p. 418).

[13] Pananides, N.A., and Arny, T. *Introductory Astronomy,* 2nd. Ed. Reading: Addison-Wesley Publishing, 1979 (p. 293).

[14] Posti, L. and Helmi, A. (2019). Mass and shape of the Milky Way's dark matter halo with globular clusters from *Gaia* and Hubble. Astronomy & Astrophysics. 621:A56. doi: 10.1051/0004-6361/201833355

[15] Tucker, W. (2006). Recent and Future Observations in the X-ray and Gamma-ray Bands: Chandra, Suzaku, GLAST, and NuSTAR, AIP Conference Proceedings. 801, 21 (arXiv:astro-ph/0512012). doi: 10.1063/1.2141828

[16] Baushev, A.N. (2012). Principal Properties of the Velocity Distribution of Dark Matter Particles Near the Solar System. Journal of Physics Conference Series. 375 012048 (doi: 10.1088/1742-6596/375/1/012048)

[17] Bowman, J.D. (2018). An Absorption Profile Centered at 78 Megahertz in the Sky-Averaged Spectrum. Nature. 555, 67. doi: 10.1038/nature25792

[18] Barkana, R. (2018). Possible interactions between baryons and dark matter. arXiv:1803.06698v1 [astro-ph.CO]

[19] McGaugh, S.S. Strong hydrogen absorption at cosmic dawn: the signature of a baryonic universe. 2018. arXiv:1803.02365v1 [astro-ph.CO] doi: 10.3847/2515-5172/aab497

[20] Read, J.I., Walker, M.G., and Steger, P. (2019). Dark matter heats up in dwarf galaxies. Mon. Not. Roy. Astro. Soc. 484: 1401–1420. arXiv:1808.06634v2 [astro-ph.CO] doi: 10.1093/mnras/sty3404

[21] Wikipedia contributors, 'Wouthuysen–Field coupling', Wikipedia, The Free Encyclopedia, 3 September 2019, 21:22 UTC, https://en.wikipedia.org/w/index.php?title=Wouthuysen%E2%80%93Field_coupling&oldid=913891210 [accessed 4 February 2020]

[22] AstroBaki. (2017). https://casper.ssl.berkeley.edu/astrobaki/index.php/Wouthuysen_Field_effect

APPENDIX

Selected Flat Space Cosmology and Related Publications 2015 thru 2020

(with online DOI links)

Tatum, E.T. (2015). Could Our Universe Have Features of a Giant Black Hole? Journal of Cosmology, 25: 13061–13080.

Tatum, E.T. (2015). How a Black Hole Universe Theory Might Resolve Some Cosmological Conundrums. Journal of Cosmology, 25: 13081–13111.

Tatum, E.T., Seshavatharam, U.V.S., and Lakshminarayana, S. (2015). The Basics of Flat Space Cosmology. International Journal of Astronomy and Astrophysics, 5: 116–124. http://doi.org/10.4236/ijaa.2015.52015

Tatum, E.T., Seshavatharam, U.V.S, and Lakshminarayana, S. (2015). Thermal Radiation Redshift in Flat Space Cosmology. Journal of Applied Physical Science International, 4(1): 18–26.

Tatum, E.T., Seshavatharam, U.V.S., and Lakshminarayana, S. (2015). Flat Space Cosmology as a Mathematical Model of Quantum Gravity or Quantum Cosmology. International Journal of Astronomy and Astrophysics, 5: 133–140. http://doi.org/10.4236/ijaa.2015.53017

Tatum, E.T., Seshavatharam, U.V.S., and Lakshminarayana, S. (2015). Flat Space Cosmology as an Alternative to ΛCDM Cosmology. Frontiers of Astronomy, Astrophysics and Cosmology, 1(2): 98–104. http://pubs.sciepub.com/faac/1/2/3

Chapter 1: Tatum, E.T. (2020). A Heuristic Model of the Evolving Universe Inspired by Hawking and Penrose. In Eugene T. Tatum (Ed.).

New Ideas Concerning Black Holes and the Universe. (pp. 5–21). London: IntechOpen. http://dx.doi.org/10.5772/intechopen.87019

Chapter 2: Tatum, E.T. (2018). Why Flat Space Cosmology Is Superior to Standard Inflationary Cosmology. Journal of Modern Physics, 9: 1867–1882. https://doi.org/10.4236/jmp.2018.910118

Chapter 3: Tatum, E.T. and Seshavatharam, U.V.S. (2018). Temperature Scaling in Flat Space Cosmology in Comparison to Standard Cosmology. Journal of Modern Physics, 9: 1404–1414. https://doi.org/10.4236/jmp.2018.97085

Chapter 4: Tatum, E.T. and Seshavatharam, U.V.S. (2018). How a Realistic Linear Rh=ct Model of Cosmology Could Present the Illusion of Late Cosmic Acceleration. Journal of Modern Physics, 9: 1397–1403. https://doi.org/10.4236/jmp.2018.97084

Chapter 5: Tatum, E.T. and Seshavatharam, U.V.S. (2018). Clues to the Fundamental Nature of Gravity, Dark Energy and Dark Matter. Journal of Modern Physics, 9: 1469–1483. https://doi.org/10.4236/jmp.2018.98091

Chapter 6: Tatum, E.T. (2018). How the CMB Anisotropy Pattern Could Be a Map of Gravitational Entropy. Journal of Modern Physics, 9: 1484–1490. https://doi.org/10.4236/jmp.2018.98092

Chapter 7: Tatum, E.T. (2018). Predicted Dark Matter Quantitation in Flat Space Cosmology. Journal of Modern Physics, 9: 1559–1563. https://doi.org/10.4236/jmp.2018.98096

Chapter 8: Tatum, E.T. (2018). Flat Space Cosmology as a Model of Penrose's Weyl Curvature Hypothesis and Gravitational Entropy. Journal of Modern Physics, 9: 1935–1940. https://doi.org/10.4236/jmp.2018.910121

Chapter 9: Tatum, E.T. (2018). Calculating Radiation Temperature Anisotropy in Flat Space Cosmology. Journal of Modern Physics, 9: 1946–1953. https://doi.org/10.4236/jmp.2018.910123

Chapter 10: Tatum, E.T. and Seshavatharam, U.V.S. (2018). Cosmic Time as an Emergent Property of Cosmic Thermodynamics. Journal of Modern Physics, 9: 1941–1945. https://doi.org/10.4236/jmp.2018.910122

Chapter 11: Tatum, E.T. and Seshavatharam, U.V.S. (2018). Flat Space Cosmology as a Model of Light Speed Cosmic Expansion—Implications for the Vacuum Energy Density. Journal of Modern Physics, 9: 2008–2020. https://doi.org/10.4236/jmp.2018.910126

Chapter 12: Tatum, E.T. (2019). My C.G.S.I.S.A.H. Theory of Dark Matter. Journal of Modern Physics, 10: 881–887. https://doi.org/10.4236/jmp.2019.108058

Chapter 13: Tatum, E.T. (2019). How the Dirac Sea Idea May Apply to a Spatially-Flat Universe Model (A Brief Review). Journal of Modern Physics, 10: 974–979. https://doi.org/10.4236/jmp.2019.108064

Chapter 14: Tatum, E.T. (2019). A Universe Comprised of 50% Matter Mass-Energy and 50% Dark Energy. Journal of Modern Physics, 10: 1144–1148. https://doi.org/10.4236/jmp.2019.109074

Chapter 15: Tatum, E.T. and Seshavatharam, U.V.S. (2020). How Flat Space Cosmology Models Dark Energy. Journal of Modern Physics, 11: 1493–1501. https://dx.doi.org/10.4236/jmp.2020.1110091

Chapter 16: Tatum, E.T. (2020). Dark Matter as Cold Atomic Hydrogen in its Lower Ground State. In Michael L. Smith (Ed.). *Cosmology 2020: The Current State.* (pp. 93–102). London: IntechOpen. https://dx.doi.org/10.5772/intechopen.91690

Appendix: Tatum, E.T. (2018). Searching for E.T. – A Universal Units Proposal. Journal of the British Interplanetary Society, 71(2): 43–44.

Searching for E.T. – A Universal Units Proposal

(lead article in Journal of the British Interplanetary Society, Feb. 2018, pp. 43–44)

Abstract: Given the geocentric nature of mks units, scientific units used by an extraterrestrial civilization are likely to be very different from our own. The following question is posed: 'Can we devise a standard set of physical units likely to be recognized by an extraterrestrial intelligent civilization?' A system of universal scientific units based upon the root mean square charge radius of the proton (L_u), rest mass of the electron (M_u) and vibration period of the super-cooled cesium atom (T_u) is proposed. This might improve recognition and understanding of any scientific numbers encoded in an intelligent signal.*

Keywords: SETI; Universal Scientific Units; Proton Charge Radius; Electron Rest Mass; Cesium Vibration Period; Mass-Energy Equivalence

INTRODUCTION AND BACKGROUND

The search for extraterrestrial intelligence has been intensified since the recent discovery of exoplanets associated with many stars examined to date. Whether Earthbound or alien, any civilization sufficiently advanced to send or receive communication signals will likely incorporate, and search for, certain number sequences. Even without the decimal point, number streams containing 3141592653… and 2718281828… should be universally recognizable as representative of natural numbers *pi* and *e*, respectively. The same is true for the dimensionless fundamental

*Originally published on July 4, 2018 in J. Brit. Interplanetary Society (see Appendix refs).

physical constants. In contrast to speed of light *c* or gravitational constant *G*, the dimensionless physical constants referred to herein contain no units. However, they are also believed to be constant over time and universal in nature. The fine-structure constant *a* is perhaps the best known of these. It can either be expressed as 0.0072973525664(17) or, more commonly, in its reciprocal form, as 137.035999139(31). Mass ratios of various Standard Model particles are also without units and considered to be constant and universal. Perhaps the most widely known of these is the proton-to-electron mass ratio *mu*, which equals 1836.15267389(17).

Scientific numbers expressed in dimensional units are another matter entirely. It is not particularly useful to encode outgoing messages from Earth containing dimensioned physical constants incorporating our standard meter, kilogram and second (mks) unit nomenclature or any other system derived and only understood in geocentric terms. Alien civilizations would surely have very different units for measuring and comparing the physical world. We cannot expect that a number stream encoded as 299792458, which we would instantly recognize as correlating to speed of light *c*, would be meaningful for any alien civilization. Rather, we must consider a unit system based upon the many things we would expect to have in common. There is considerable evidence that the laws of physics are the same everywhere we look in the universe. It is, therefore, likely that intelligent extraterrestrial life, if it does exist, would have discovered many of the same laws of nature that we have. Moreover, they may already have established a more logical set of physical units than our parochial and geocentric mks system.

A UNIVERSAL UNITS PROPOSAL

As for the problem of incorporating and searching for the numerous scientific numbers requiring units, it is the purpose of this paper to propose a rational solution to the following question:

'Can we devise a standard set of physical units likely to be recognized by an extraterrestrial intelligent civilization?'

If we were to start from scratch, we might choose units of length, inertia and time that would be easily discoverable anywhere in the universe. Since the visible universe is largely composed of atomic matter, a reasonable place to start in devising a logic-based set of universal physical units is with subatomic quanta which could be easily discovered by any intelligent civilization. In particular, quanta of the most common and basic universal element, hydrogen, would be reasonable and likely candidates. If we look back at our own history of early discovery, there was particular focus on this simplest of elements. It is not accidental that the electron and, soon after, the proton were the first subatomic particles discovered and characterized. Given their ease of discovery and measurement and their obvious unitary nature, it is proposed that they should be the basis for unitary charge, mass and length standardization. Although one could argue, at least in the case of the proton, that quarks and gluons are even more basic, their precise characteristics have been very difficult to establish, and nature doesn't appear to allow for experimentation with individual quarks. They appear to be perpetually locked within the proton and neutron.

Of the two above-mentioned subatomic quanta for consideration, the electron has a rest mass M_e considerably smaller than that of the proton. Therefore, M_e is chosen as the likely candidate for a universal mass unit. In our own mks system, M_e is $9.10938356(11) \times 10^{-31}$ kg. Thus, the kilogram can also be considered to be equal to $1.0977691227 \times 10^{30}$ electron rest mass units. A length unit, of course, cannot be established from the electron, since it behaves as a point particle with no measurable spatial dimension. Therefore, a logical choice for a fundamental universal length unit is the root mean square charge radius of the proton, which is $8.751(61) \times 10^{-16}$ m [1]. Thus, the meter can also be considered to be equal to 1.14265×10^{15} proton radii.

Establishing a time interval as a likely candidate for a universal time unit is a somewhat trickier proposition. Obviously, in keeping with the above theme, some regular periodic atomic behavior should be chosen as the time standard. Subatomic particle spins cannot be chosen, because "spin" in this context does not mean actual regular particle rotation, but rather something more nebulous. Therefore, atomic energy level transitions of characteristic photon wavelength and frequency become a logical next place to look for a fundamental vibration period. In our Earth-based system, a great deal of time and effort has already been given to this question. The logical and practical choice for a number of years has been an energy transition within the super-cooled cesium atom with a frequency of 9.19263177×10^{9} Hz and a period of $1.087827757 \times 10^{-10}$ s. And although there are a number of competing choices for regular atomic vibration, this cesium choice appears to be especially regular and reliable and, therefore, a likely candidate for a universal time unit. This proposed cesium time interval standard is reported to be accurate to within one second per roughly 30 million years.

A secondary advantage of the above cesium time standard as a universal time unit candidate is the alternative universal length standard it implies with respect to speed of light c. Since length is the product of velocity and time, this cesium time standard implies that light in a vacuum travels 2.99792458×10^8 m/s times $1.087827757 \times 10^{-10}$ s $= 3.2612255715 \times 10^{-2}$ m (call it the 'cesium length') per cesium vibration period (call it the 'cesium time'). This distance of 3.2612255715 cm (approximately 1.284 inch) could be a physically useful alternative length standard in the everyday world of human or like-sized alien. Also, such a system of cesium time and length standards allows for speed of light c a normalized value of unity (one length unit per time unit). This can be useful in the many physics equations incorporating c. To give but one example, *in terms of the unit coefficients, $E = mc^2$ simply reduces to $E = m$ in a universal system quantifying the speed of light in terms of cesium time and cesium length standards. Changes in the potential and kinetic energies of any material object increase or decrease inertia m by an equivalent amount because the unit coefficient is the same number. The energy-mass equivalence discovered by Einstein perhaps becomes a little less mysterious in a world where the coefficients for the energy units and the mass units are identical.*

Staying with the proton charge radius (L_u), electron rest mass (M_u) and cesium vibration period (T_u) as respective universal length, mass and time unit alternatives to our geocentric mks system, one can easily calculate speed of light c and gravitational constant G in these new units. Referring to **Table 1**, c becomes $3.7264 \times 10^{13}\ L_u/T_u$ and G becomes $1.073348 \times 10^{-15}\ L_u^3/M_uT_u^2$. The Planck constant h becomes $1.033267758 \times 10^{17}\ M_uL_u^2/T_u$ and the reduced Planck constant $\hbar$ becomes $1.6444967122 \times 10^{16}\ M_uL_u^2/T_u$. If one chooses to work with Planck length ($\hbar^{1/2}G^{1/2}c^{-3/2}$), Planck time ($\hbar^{1/2}G^{1/2}c^{-5/2}$) and Planck mass ($\hbar^{1/2}c^{1/2}G^{-1/2}$) in this proposed universal unit system, one can easily convert these Planck formulae into the appropriate universal numbers and units by making use of the converted values of c, G, h and $\hbar$ provided in **Table 1**.

Table 1 Proposed universal units and representative constants.

Length (L_u)	Mass (M_u)	Time (T_u)
Proton Radius	**Electron Mass**	**Ceisum Period**
$8.751(61) \times 10^{-16}$ m	$9.10938356(11) \times 10^{-31}$ kg	$1.087827757 \times 10^{-10}$ s
$1.14265 \times 10^{15}\ L_u$/m	$1.0977691227 \times 10^{30}\ M_u$/kg	$9.19263177 \times 10^{9}\ T_u$/s
	$c = 3.7264 \times 10^{13}\ L_u/T_u$	
	$G = 1.073348 \times 10^{-15}\ L_u^3/M_uT_u^2$	
	$h = 1.033267758 \times 10^{17}\ M_uL_u^2/T_u$	
	$\hbar = 1.6444967122 \times 10^{16}\ M_uL_u^2/T_u$	

The resulting converted Planck length, time and mass values (not shown) can also be incorporated and searched for as number sequences in signals communicating the presence of intelligent life.

The various number sequences given or alluded to in this paper have so far been in the decimal system format. Humans have presumably evolved such a system based on the total number of digits on the upper extremities. However, if one considers the number variety of extremity digits seen in the wide spectrum of terrestrial species, it should be obvious that all number sequences given or alluded to in this paper should also be incorporated and searched for in translated form within modular numbering systems other than the decimal system.

SUMMARY

A system of universal scientific units based upon the root mean square charge radius of the proton (L_u), rest mass of the electron (M_u) and vibration period of the super-cooled cesium atom (T_u) is proposed. This might improve recognition and understanding of any scientific numbers encoded in an intelligent signal, should we intercept one.

REFERENCE

[1] Mohr, P.J., Newell, D.B. and Taylor, B.N. CODATA Recommended Values of the Fundamental Physical Constants: 2014. arXiv:1507.07956 [physics.atom-ph] doi:10.1103/RevModPhys.88.035009

Dedications and Acknowledgements

Both authors dedicate this book to the late Dr. Stephen Hawking and to Dr. Roger Penrose for their groundbreaking work on black holes and their possible application to cosmology. Dr. Tatum thanks Dr. Rudolph Schild of the Harvard-Smithsonian Center for Astrophysics for his past encouragement and support. Author Seshavatharam is indebted to professors Brahmashri M. NagaphaniSarma, Chairman, Shri K.V. Krishna Murthy, founding Chairman, Institute of Scientific Research in Vedas (I-SERVE), Hyderabad, India, Shri K.V.R.S. Murthy, former scientist IICT (CSIR), Govt. of India, Director, Research and Development, I-SERVE, and Prof. S. Lakshminarayana, Dept. of Nuclear Physics, Andhra University, Visakhapatnam, India, for their valuable guidance and great support in developing this subject.

About the Authors

Eugene Terry Tatum is today's foremost expert on the apparent links between black holes and cosmology. This field of interest stems from the original work of Penrose and Hawking in the 1960s and continues today with the Flat Space Cosmology model summarized herein. Dr. Tatum has numerous peer-reviewed physics journal publications and was the editor of IntechOpen's 2020 book entitled *New Ideas Concerning Black Holes and the Universe.* He is also the originator of the 'cold hydrogen dark matter' theory presented herein.

U.V.S. Seshavatharam is a member of I-SERVE, Hyderabad, India. He has accumulated more than one hundred publications in numerous peer-reviewed physics journals. Most of these have been in co-authorship with Prof. S. Lakshminarayana, formerly of the Dept. of Nuclear Physics, Andhra University, AP, India. Current theoretical interests include quantum gravity, quantum cosmology and cold nuclear fusion.

Book Synopsis

This compilation based upon recent peer-reviewed journal publications encapsulates how the Flat Space Cosmology model (FSC) has become the primary competitor to the inflationary standard model of cosmology. New ideas concerning black holes, dark energy and dark matter are presented and shown to correlate extremely well with astronomical observations.

Anyone who follows the fast-changing science of cosmology, has an interest in the latest developments, and would like to know how it is that our universe appears to follow equations one would ordinarily expect for a time-reversed black hole (!), may find this book to be fascinating.

Cosmology is the study of how the universe has changed over the great span of time (roughly 14 billion years). Later centuries will look back upon the period from 1990–2030 as a 'Golden Age' of theoretical and observational cosmology. It is highly likely that we are on the verge of a deeper understanding of the most mysterious energy ('dark energy') and matter ('dark matter') comprising the *majority* of energy and matter in the universe. Some of the material presented in this book is on the cutting edge of dark energy and dark matter theoretical work.

This book summarizes, for the first time, the groundbreaking publications of two cosmologists, one from the United States and the other from India, from 2015 thru 2020. During this highly productive period, the authors stealthily published their papers in six different peer-reviewed scientific journals, so that the model could be quietly explored in all aspects before bringing it all together in a single book. This is that book!

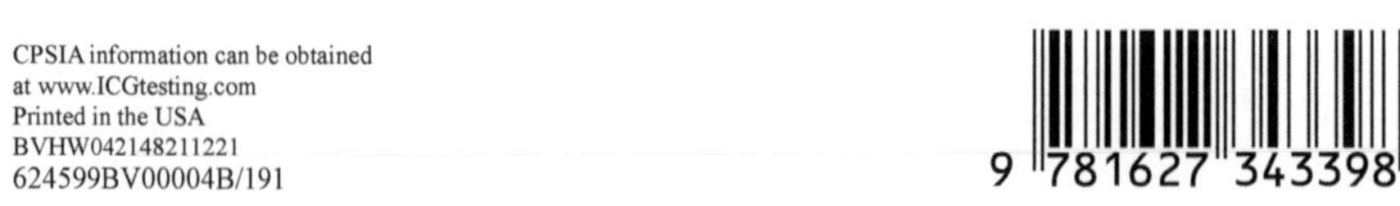

CPSIA information can be obtained
at www.ICGtesting.com
Printed in the USA
BVHW042148211221
624599BV00004B/191

9 781627 343398